TESLA

Man Out of Time

Margaret Cheney

A Touchstone Book
Published by Simon & Schuster
New York   London   Toronto   Sydney

The airplane on the facing page, patented in the 1920s by Tesla, was intended to operate much like the vertical/short takeoff and landing (V/STOL) aircraft being considered in the 1980s by the U.S. Navy as "a subsonic aircraft for the 1990s." The latter is a jet aircraft with sophisticated electronic equipment, but the basic concept is the same. Tesla's plane was never built.

🌿 TOUCHSTONE
A Division of Simon & Schuster, Inc.
1230 Avenue of the Americas
New York, NY 10020

First Touchstone edition 2001

TOUCHSTONE and colophon are registered trademarks of Simon & Schuster, Inc.

For information about special discounts for bulk purchases, please contact Simon & Schuster Special Sales at 1-866-506-1949 or business@simonandschuster.com.

Manufactured in the United States of America

30 29

Library of Congress Cataloging-in-Publication Data
Cheney, Margaret.
    Tesla : man out of time / Margaret Cheney. First Touchstone ed.
        p. cm.
    Includes bibliographical references and index.
    1. Tesla, Nikola, 1856–1943.   2. Electrical engineers—United States—Biography.   3. Inventors—United States—Biography.
TK140.T4 C47 2001
621.3'092—dc21
[B]   2001037808
ISBN 978-0-7432-1536-7

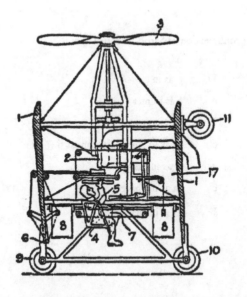

For Barbara Nelson and Allen Davidson

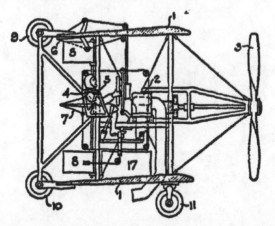

To aid in the pronunciation of Serbo-Croatian names:

Đ	g as in gin
e	e as in met
ž	z as in azure
ć	ch as in chin; soft, almost tj
c	ts as in fits
č	ch as in charge
š	sh as in shall

CONTENTS

ACKNOWLEDGMENTS

I wish particularly to thank:

Leland Anderson, one of the founders of the Tesla Society,* a coauthor of the annotated *Dr. Nikola Tesla Bibliography* (San Carlos, Ca., Ragusan Press, 1979), and author of the monograph, "Priority in the Invention of Radio—Tesla v. Marconi." Mr. Anderson's research and scholarly works on Tesla have been a major interest of his life. An electrical engineer and former computer consultant, he reviewed my manuscript and generously shared his collection of Tesliana, including many previously unpublished materials and photographs.

Maurice Stahl, a physicist formerly with the Hoover Company and now a consultant for the McKinley Historical Museum in Ohio (featuring a Tesla exhibit), also reviewed the manuscript and advised on technical aspects.

Dr. Bogdan Raditsa, who served under President Tito of Yugoslavia in the early days of his administration, clarified and amplified Yugoslav-Allied politics during World War II as it affected Tesla. He has lived in America for many years, writes books and articles, and teaches Balkan history at Fairleigh Dickinson University.

Dr. Lauriston S. Taylor, a radiological physics consultant and recent past president of the NCRP, is an authority on the pioneers of X ray and, as such, read and commented on Tesla's contributions in this field.

Lambert Dolphin, assistant director of the Radio Physics Laboratory, SRI International, analyzed Tesla's research in ball lightning, particle-beam weapons, radio communication, and alternating current.

Dr. James R. Wait, formerly senior scientist at the National Oceanic

---

* The society has been disbanded.

and Atmospheric Administration environmental research laboratories at Boulder, Colorado, and an authority on wave propagation, commented on Tesla's concept of electromagnetic energy being transmitted "through the Earth," as did Mr. Anderson.

Professor Warren D. Rice of Arizona State University, a leading researcher on the Tesla turbine, also analyzed Tesla's theories of terrestrial-heat power plants and ocean thermal energy conversion plants against the foreground of contemporary work.

I am indebted to radio pioneer Commander E. J. Quinby (USN Ret.), who contributed his personal reminiscences of Tesla's early work in radio and robotry; to Dr. Albert J. Phillips, former research director of ASARCO, for memories of working with Tesla on a research project; and to Dr. William M. Mueller, Colorado School of Mines, Department of Metallurgy, who gave his analysis of the ASARCO experiment.

Of Tesla's many loyal admirers in America, few have worked as tirelessly to see justice done to his memory as Nick Basura. He guided me to useful sources at the beginning of my research, for which I am grateful.

Harry Goldman, a Tesla scholar and writer-photographer, provided valuable information and special services in the enhancing of old photographic prints, as well as contributing photos from his private collection.

I am grateful to Eleanor Treibek of Volunteers in Action, the Language Bank, Monterey, California, for translation assistance; to the Tesla Museum in Belgrade, Yugoslavia, for photographs and for the letters of Katharine Johnson, and of Michael Pupin, A. J. Fleming, Sir William Crookes, Richmond P. Hobson, and other tributes to Nikola Tesla; to Professor Philip S. Callahan for permission to use his photograph of Tesla's birthplace; to the Butler Library at Columbia University for photographs and the letters of Robert and Katharine Johnson, George Scherff, Nikola Tesla, George Westinghouse, Major Edwin Armstrong, and Leland Anderson; to the Manuscript Division of the Library of Congress for the microfilm letters of Nikola Tesla, Robert Johnson, Mark Twain, B. F. Meissner, George Scherff, George Westinghouse, J. Pierpont Morgan, J. P. Morgan, and others; to Archivist J. R. K. Kantor of the Bancroft Library, University of California, Berkeley, for access to the Julian Hawthorne Papers, and to the university's History of Science and Technology Project; to the reference staffs of the John Steinbeck Library at Salinas, the New York Public Library, and the libraries of the Massachusetts Institute of Technology; to the librarian of Special Collections, Purdue University; to Mr. Elliot N. Sivowitch and the Smithsonian Institution, National Museum of American History, to the Westinghouse Corporation, the Brookhaven National Laboratory, RCA, and Niagara Mohawk—for photographs; to Robert Golka for information on "Project Tesla."

I wish also to thank the Federal Bureau of Investigation, the Department of the Navy, the National Security Agency, the Central Intelligence Agency, the National Archives and Record Service, the technical librarian of Wright-Patterson Air Force Base, the Office of Alien Property, and the Office of the Medical Examiner, City of New York.

The author and the publisher also gratefully acknowledge indebtedness for quotations in the text of this book as follows:

To Dr. Jule Eisenbud and Mrs. Laura A. Dale for permission to quote from the article, "Two Approaches to Spontaneous Case Material," by Eisenbud, in the *Journal of the American Society for Psychical Research* of July 1963; to the David McKay Company for permission to quote from John J. O'Neill's *Prodigal Genius* (originally published by Ives Washburn, Inc., 1944); to *The New York Times* for lines from "Electrical Sorcerer," by Waldemar Kaempffert, Book Review Section, Feb. 4, 1945; to *Time* magazine for lines from its cover story on Nikola Tesla of July 20, 1931; to Frederic B. Jueneman for permission to quote from *Limits of Uncertainty*, p. 206f, Dun-Donnelly, Chicago, 1975; and to Jueneman and *Industrial Research* to quote from "Innovative Notebook," by Jueneman, February 1974.

Thanks to *Science & Mechanics* for permission to quote from the article, "Our Future Motive Power," by Nikola Tesla, *Everyday Science & Mechanics*, December 1931, and to reproduce an illustration therefrom.

Especially thanks to M. Harvey Gernsback, president of Gernsback Publications, Inc., for permission to reprint photos, illustrations of the artist Frank Paul, and quotes from "My Inventions," by Nikola Tesla, that appeared in the *Electrical Experimenter* and *Science & Invention*, formerly published by Hugo Gernsback.

And to Leland Anderson for permission to quote from "Priority in Invention of Radio, *Tesla v. Marconi*," Antique Wireless Association, March 1980.

In addition the author is indebted to the Nikola Tesla Museum for words quoted from *Colorado Springs Notes, 1899–1900,* by Nikola Tesla; to King Peter II for a quotation from *A King's Heritage*, Putnam, New York, 1954, and for lines from T. C. Martin, ed., *The Inventions, Researches and Writings of Nikola Tesla*, reprinted from *The Electrical Engineer*, 1894 (reissued by Omni Publications, Hawthorne, Calif., 1977).

Among friends and relatives, I am grateful to inventor Allen Davidson and to Randy Pierce and "PJ," who bravely read and commented on the manuscript in its earliest phases and heartened me with their enthusiasm. Most of all I thank Barbara Nelson for her editorial criticism and loyalty throughout a lengthy endeavor.

PREFACE

It is astounding how time, neglect, and a dazzling new generation of technocrats have converged upon the patents and reputation of this nineteenth-century genius. Nikola Tesla's work, although sometimes pirated or forgotten and often misunderstood, has never really needed burnishing. One suspects that he is chuckling now as the world catches up to his visions and concepts since, as he wisely noted long ago, "The scientific man does not aim at an immediate result. He does not expect that his advanced ideas will be readily taken up. His work is like that of a planter—for the future."

In a mere decade this planter's once almost forgotten name has acquired a new aura of symbolic allure that feels more at home in an era of automation and the internet. It speaks somehow to the great wealth that was never to be his, to the billions of dollars being earned or anticipated by his smart young admirers of the twenty-first century who have often commented that they are inspired by his work, by his patents, and notably by his dreams. A brilliant neurotic, Tesla spent his last years as an octogenarian in a New York hotel room subsisting on a diet of Nabisco crackers and the spiritual companionship of a lovely white pigeon.

Today one can scarcely pick up a copy of *The Wall Street Journal* or *The New York Times* without finding a mention of him or his effect on famous young followers and admirers.

It is relevant to mention that he died alone and in debt in 1943, whereupon busy, prodigal America, his chosen country, simply forgot about him in the aftermath of World War II. And as if that were not enough, the academic official assigned to evaluate and preserve his scientific estate—all the research and records he had amassed in more than fifty years—referred them to the Office of Alien Property Custodian (though Tesla was a naturalized U.S. citizen!), which in turn released them to his homeland, the small country of Serbia (then part of communist Yugoslavia). Immediately, of course, this intellectual trove was declared off-limits for generations of Americans and all other citizens except possibly Russians.

One may well ask, is this a hard-luck story?

When I first heard the name Tesla, surprisingly he was known to relatively few Americans, mention of him being almost completely missing from science textbooks through the university level. Even his fellow electricians and engineers were often mystified by his boasts of taming the wild kinetic energy of earth and sky. (Indeed, had it not been for a certain ham radio operator named John Wagner, a Michigan third-grade schoolteacher who correctly protested that it had been Tesla and not the Italian fellow Guglielmo Marconi who invented radio, and who, with his students and the rock group TESLA, devoted decades of energy to clearing away misconceptions—even those of the Smithsonian National Museum of Natural History, which credited Edison and Marconi with some of Tesla's inventions—the truth would have taken even longer to emerge.) Of course there were scientists who knew or suspected that Tesla deserved the credit; but the Nobel Prize Committee awarded the physics prize to Marconi for inventing wireless telegraphy and was loath to make a correction, which incidentally it has not done to this day. In 1943 the U.S. Supreme Court decided in Tesla's favor, but who kept count during a world war? And unfortunately, the inventor had died.

An uneasy relationship between Tesla and Thomas Edison was definitively poisoned in the early twentieth century by a cutthroat affair known as the War of the Currents between their respective industrial champions, the Westinghouse Corporation and General Electric Company. This was resolved with the triumph of Tesla's alternating current system in harnessing Niagara Falls, and ultimately with AC electrification of the entire world. Alternating current, augmented by radio and automation, has revolutionized the basic technologies of twenty-first-century progress.

In the marshes of public perception, however, it was the all-American Edison (who gave us the phonograph, the lightbulb, and waxed paper, among many other sensible items) who for generations sprang to mind in debates over the major achievements of the twentieth century. Many of Edison's triumphs, unfortunately, have recently faded from everyday use, including most recently the incandescent lightbulb. In Europe a more efficient light is being developed, a descendant of the fluorescent tube that Tesla first demonstrated at the Chicago World's Fair in 1893.

The rediscovery of Nikola Tesla (not to mention of his great work) has influenced some of the twenty-first century's important pioneers. Larry Page, who with his partner Sergey Brin created the global search engine Google, says that his interest in science started early. As explained in *The Economist*, "When he was twelve, he read a biography of Nikola Tesla, a prolific inventor who never got credit for much, but is now a hero among geeks." Page, a bona fide visionary, considered writing his thesis at Stanford University on Tesla but decided to "play it safe," as he said, and proceeded to invent Google.

When a California company recently announced its plans to produce a high-speed all-electric sports car, the entrepreneurs unsurprisingly chose the name Tesla Motor Company. Hollywood admirers predictably rushed to invest millions in its future. Almost a century earlier, Tesla (one of the earliest environmentalists) had experimented with all-electric cars with limited success. He had also designed a small plane called a "flying stove" that was meant to carry the commuter from an upstairs bedroom window directly to his office. It remains one of several unfulfilled dreams.

The eccentric and controversial genius who has emerged from obscurity to become in our time the inspiration of billionaire geeks,

New Age spiritualists, rock stars, third graders, schoolteachers, poets, operatic composers, playwrights, movie and TV producers, artists, photographers, and Dadaist sculptors, as well as the subject of a stream of books both scholastic and fantastic, may be one of the truly restorative figures of our age.

Tesla mythology is commercially employed today by entrepreneurs as varied as chip manufacturers and the makers of video games. The marketers of a new product line at Nvidia compete with chip competitors by selling a line of Tesla products. It was no doubt inevitable that the builders of a video game named Dark Void would center their conspiracies on a character named Nikola Tesla.

It is scarcely surprising to learn that some of Tesla's old patents are providing the basic concepts used in modern atmospheric physics. A controversial military-industrial project known as HAARP, the high frequency active auroral research program, is a many-pronged exploration of the ionosphere based in Alaska. The Russian Duma has protested that it will infringe on world communication systems. Technically the earth's ionosphere was not discovered until 1926 by British physicist E. V. Appleton. Yet in 1900 Tesla filed a patent for the wireless transmission of energy through this still nameless region some seventy kilometers above earth, where particles of cosmic matter from the sun are trapped between the vacuum of space and earth's atmosphere.

Tesla was galvanized by the luminous mass of swirling energy known as the northern lights (aurora borealis) that create a huge electrodynamic circuit above the earth, carrying the energy of a thousand large power plants. No one can be certain where this research will lead. Excitable writers warn that HAARP'S "death rays" may set the sky ablaze (which is ridiculed by its creators) or permit universal mind control. Others speak of world weather control (a mixed blessing on a fractious planet) and of the ability to transmit energy by microwaves around the globe.

Tesla's basic idea was revisited by Atlantic Richfield Oil Company's consultant physicist Bernard Eastlund, who sought a less expensive way to transmit natural gas from remote areas. He proposed a "pipeline" in the sky, using natural gas to power microwave transmitters. The microwaves would be sped great distances through the ionosphere, then beamed down to satellites and converted to electrical power on earth. In short, HAARP is simply vintage Tesla in modern hands, another reminder that he still looks down upon us.

M.C.
2010

FOREWORD
Leland Anderson

Despite the flashy, dramatic, and often limelight attention that Nikola Tesla was given in the heyday of his reign in the fields of research and engineering, he maintained a very private personal life. Since he was a loner—a perennial bachelor, working apart, not entering into corporate associations, and not mixing friends—his personal life was obscure to outsiders. Such reclusiveness marking the career of one of the world's leading figures in science and engineering can pose severe analytical obstacles for a biographer. However, almost immediately after Tesla's death at the age of eighty-six in 1943, the biography *Prodigal Genius* appeared by John J. O'Neill, science editor of the New York *Herald Tribune*. For many years it stood as the only biography of Tesla, primarily because of the difficulty for any other would-be biographer to uncover significant additional information about him.

Following World War II, the tons of material representing Tesla's library were shipped to Belgrade, Yugoslavia, the country of his birth (Tesla was a U.S. citizen), where a state museum was established in his name. The circumstances surrounding the transfer of his estate to Yugoslavia are interesting but will not be commented upon here except to point out the problem of remoteness of such a museum for any biographer in this country, let alone the severe restrictions on access to archival materials that exist for researchers venturing to the museum.

In 1959, two rather short biographies of Tesla appeared. Dr. Helen Walter's book was intended for young people, and curiously contained illustration and frontispiece sketches quite unlike Tesla's appearance.

Margaret Storm's book, published by herself and printed in green ink, was based on the assertion that Tesla was an embodiment of a superior being from the planet Venus! Another short biography intended for young people appeared in 1961 by Arthur Beckhard. Tesla's name was misspelled on the dust jacket (Tesla once wrote to a friend that he wished he could turn all the forked lightning in his laboratory on critics who misspell his name), and the book omits essentially everything on his life after 1900 (Tesla was then forty-four). All three authors leaned heavily on O'Neill's biography, as evidenced by the perpetuation of a number of erroneous legends that subsequent study has vitiated, and none of the three extended O'Neill's treatment.

Lightning in His Hand: The Life Story of Nikola Tesla, by Inez Hunt and Wanetta Draper, nearby residents of Colorado Springs, appeared in 1964, twenty years after O'Neill's biography. O'Neill did not venture to Colorado Springs, where Tesla established an experimental station in 1899 and conducted electrical experiments which to this very day amaze scientists the world over, and consequently did not benefit from information that could have been provided by residents of that city about Tesla's interactions with them. Tesla took on flesh and bones to some degree in Hunt and Draper's biography, and the book carried numerous photographs. Much of the focus of the book concerned Tesla's half-year stay in the Springs, which was the original intent of the authors.

Why should anyone actually wish to undertake another full biography after the appearance of O'Neill's *Prodigal Genius*? It has been considered the most authoritative biography extant, and probably was the best effort that could have been produced by anyone at that time, with the exception of Kenneth Swezey—a science writer and Tesla's close personal friend during the last twenty-plus years of his life. However, from this vantage point of distance in time, O'Neill's biography is now seen to be weak insofar as it analyzed Tesla the man and thin with regard to his interactions with personal associates and friends. Even though O'Neill and Tesla were amicable, Tesla kept

O'Neill at a distance, and O'Neill gleaned only what he was able to pry out of Tesla with great difficulty—certainly not the most ideal liaison for a biographer.

Much information has surfaced since the appearance of O'Neill's biography, adding new dimensions to the extent of knowledge about Tesla. Many questions asked by students of his life have been answered; however, this unfolding has also presented many more mysteries. The Freedom of Information Acts revealed that the federal government had a great interest in Tesla's papers. Why shouldn't it? In the midst of World War II, and at press conferences, Tesla often startled reporters with talk of developing weapons with beams that would melt aircraft, telegeodynamics, and other advanced concepts. Whether real or speculative, the federal government took no chances. What became of these investigations by federal agencies is a story in itself.

In reviewing my own interest in Tesla, since high school days I was fascinated by his high frequency, high voltage researches for which he became world known. I was disturbed, however, by the inordinate difficulty in obtaining copies of his technical writings and, as well, identifying references to writings by others about Tesla's work. This prompted what was to become a project of many years—that of producing an exhaustive catalog (published in 1979 as a bibliography and for which I served as co-editor) of the writings by and about Tesla and his work. In the course of pursuing studies in electrical engineering, and continuing interest in Tesla's high frequency, high voltage researches, my inquiries eventually led me to meet those who worked for him, such as his secretaries Dorothy Skerritt and Muriel Arbus, and laboratory technicians such as Walter Wilhelm. Along the way, his personal friends came into the picture as well as others who had known Tesla on a person-to-person basis.

As the Tesla Centennial (1956) approached, it became apparent that no observances were being arranged by the major scientific and engineering organizations in this country to signal the event. Together with Skerritt, Arbus, Wilhelm, and a number of other interested persons, therefore, I helped found the Tesla Society—the function of which was to develop and coordinate activities for the centennial observance. Following the centennial year, the Society expired, but an awareness of Tesla's impact on society was regenerated in the hiatus since his death. An interest had been reawakened in the discoveries that he announced and demonstrated, but which had been retarded in development because of a technology lag in associated disciplines, such as material sciences.

Inspiration—that is what he gave to other inventors whose endeavors his life spanned, and that is what his work continues to give to technical specialists in these times. On the occasion of Tesla's seventy-fifth birthday (1931), his contemporaries wrote that his lectures were then both as imaginative and inspirational to productive development as when they were first published forty years before that:

In almost every step of progress in electrical power engineering, as well as in radio, we can trace the spark of thought back to Nikola Tesla. There are few indeed who in their lifetime see realization of such a far-flung imagination. *(E. F. W. Alexanderson)*

In reading of Tesla's work one is constantly struck by his many suggestions which have anticipated later developments in the radio art. *(Louis Cohen)*

Prolific inventor, who solved the greatest problem in electrical engineering of his time, and gave to the world the polyphase motor and system of distribution, revolutionizing the power art and founding its phenomenal development. My contact as your assistant at the historic Columbia University high frequency lecture and afterward has left an indelible impression and inspiration which has influenced my life. *(Gano Dunn)*

You fanned into a never dying flame my latent interest in gaseous conduction. Early in 1894 I told our mutual friend that your book . . . which contains your original lectures, would still be considered a classic a hundred years hence. I have not changed my opinion. *(D. McFarlan Moore)*

I remember vividly the eagerness and fascination with which I read your account of the high tension experiments more than forty years ago. They were most original and daring: they opened up new vistas for exploration by thought and experiment. *(W. H. Bragg)*

There are three aspects of Tesla's work which particularly deserve our admiration: The importance of the achievements in themselves, as judged by their practical bearing; the logical clearness and purity of thought, with which the arguments are pursued and new results obtained; the vision and the inspiration, I should almost say the courage, of seeing remote things far ahead and so opening up new avenues to mankind. *(I. C. M. Brentano)*

Today, we yet find that the writings of Tesla retain their undiminished power of inspirational endeavor to the reader. Tesla was indeed *out of his time*, and this biography represents a distinct achievement in overcoming unusual investigative obstacles to bring his remarkable story to life.

Denver, Colorado

1. MODERN PROMETHEUS

Promptly at eight o'clock a patrician figure in his thirties was shown to his regular table in the Palm Room of the Waldorf-Astoria Hotel. Tall and slender, elegantly attired, he was the cynosure of all eyes, though most diners, mindful of the celebrated inventor's need for privacy, pretended not to stare.

Eighteen clean linen napkins were stacked as usual at his place. Nikola Tesla could no more have said why he favored numbers divisible by three than why he had a morbid fear of germs or, for that matter, why he was beset by any of the multitude of other strange obsessions that plagued his life.

Abstractedly he began to polish the already sparkling silver and crystal, taking up and discarding one square of linen after another until a small starched mountain had risen on the serving table. Then, as each dish arrived, he compulsively calculated its cubic contents before lifting a bite to his lips. Otherwise there could be no joy in eating.

Those who came to the Palm Room for the express purpose of observing the inventor might have noted that he did not order his meal from the menu. As usual, it had been specially prepared beforehand according to his telephoned instructions and now was being served at his request not by a waiter but by the maître d'hôtel himself.[1]

While Tesla picked at his food, William K. Vanderbilt paused to chide the young Serb for not making better use of the Vanderbilt box at the opera. And shortly after he left, a scholarly-looking man in a Van Dyke beard and small rimless glasses came to Tesla's table and greeted him with particular affection. Robert Underwood Johnson, in addition to being a magazine editor and poet, was a socially ambitious and well-connected bon vivant.

Grinning, Johnson bent down and whispered in Tesla's ear the latest rumor circulating among the "400": a demure schoolgirl named Anne Morgan, it seemed, had a crush on the inventor and was pestering her papa, J. Pierpont, for an introduction.

Tesla smiled in his modest way and inquired after Johnson's wife, Katharine.

"Kate has asked me to bring you to lunch on Saturday," said Johnson.

They discussed for a moment another guest of whom Tesla was fond—but only in a platonic way—a charming young pianist named Marguerite Merington. Assured that she too had been asked, he accepted the invitation.

The editor went his way, and Tesla returned his attention to the cubic contents of his dessert course. He had barely completed his calculations when a messenger appeared at his table and handed him a note. He recognized at once the bold scrawl of his friend Mark Twain.

"If you do not have more exciting plans for the evening," wrote the humorist, "perhaps you will join me at the Players' Club."[2]

Tesla scribbled a hasty reply: "Alas, I must work. But if you will join me in my laboratory at midnight, I think I can promise you some good entertainment."

It was, as usual, precisely ten o'clock when Tesla rose from his table and vanished into the erratically lighted streets of Manhattan.

Strolling back toward his laboratory, he turned into a small park and whistled softly. From high in the walls of a nearby building came a rustling of wings. Soon a familiar white shape fluttered to rest on his shoulder. Tesla took a bag of grain from his pocket, fed the pigeon from his hand, then wafted her into the night, and blew her a kiss.

Now he considered his next move. If he continued on around the block, he would feel compelled to circle it three times. With a sigh, he turned and walked toward his laboratory at 33–35 South Fifth Avenue (now West Broadway), near Bleecker Street.

Entering the familiar loft building in the darkness, he closed a master switch. Tube lighting on the walls sprang into brilliance, illuminating a shadowy cavern filled with weirdly shaped machinery. The strange thing about this tube lighting was that it had no connections to the loops of electrical wiring around the ceiling. Indeed, it had no connections at all, drawing all its energy from an ambient force field. He could pick up an unattached light and move it freely to any part of the workshop.

In a corner an odd contraption began to vibrate silently. Tesla's eyes narrowed with satisfaction. Here under a kind of platform, the tiniest of oscillators was at work. Only he knew its awesome power.

Thoughtfully he glanced through a window to the black shapes of tenements below. His hardworking immigrant neighbors appeared safely asleep. The police had warned him of complaints about the blue lightning flaring from his windows and electricity snapping through the streets after dark.

He shrugged and turned to his work, making a series of microscopic adjustments to a machine. Deep in concentration, he was unaware of the passage of time until he heard a pounding on the door at street level.

Tesla hurried down to greet an English journalist, Chauncey McGovern of *Pearson's Magazine.*

"I'm so pleased you could come, Mr. McGovern."

"I felt I owed it to my readers, sir. Everyone in London is talking about the New Wizard of the West—and they don't mean Mr. Edison."

"Well, come along up. Let's see if I can justify my reputation."

As they turned to the stairs there came a ring of laughter from the street entrance and a voice that Tesla recognized.

"Ah, that's Mark."

He opened the door again to welcome Twain and the actor Joseph Jefferson. Both had come directly from the Players' Club. Twain's eyes sparkled in anticipation.

"Let's have the show, Tesla. You know what I always say."

"No, what do you say, Mark?" the inventor asked with a smile.

"What I always say, and mind you they'll be quoting me into the hereafter, is that thunder is good, thunder is impressive, but it is lightning that does the work."

"Then we'll get a storm of work done tonight, my friend. Come along."

"Not to stagger on being shown through the laboratory of Nikola Tesla," McGovern would later recall, "requires the possession of an uncommonly sturdy mind. . . .

"Fancy yourself seated in a large, well-lighted room, with mountains of curious-looking machinery on all sides. A tall, thin young man walks up to you, and by merely snapping his fingers creates instantaneously a ball of leaping red flame, and holds it calmly in his hands. As you gaze you are surprised to see it does not burn his fingers. He lets it fall upon his clothing, on his hair, into your lap, and, finally, puts the ball of flame into a wooden box. You are amazed to see that nowhere does the flame leave the slightest trace, and you rub your eyes to make sure you are not asleep." [3]

If McGovern was baffled by Tesla's fireball, he was at least not alone. None of his contemporaries could explain how Tesla produced this oft-repeated effect, and no one can explain it today.

The odd flame having been extinguished as mysteriously as it appeared, Tesla switched off the lights, and the room became black as a cave.

"Now, my friends, I will make for you some daylight."

Suddenly, the whole laboratory was flooded with strange, beautiful light. McGovern, Twain, and Jefferson cast their eyes around the room, but they could find no trace of the source of the illumination. McGovern wondered vaguely if this eerie effect might somehow be connected with a demonstration Tesla had reportedly given in Paris in which he had produced illumination between two large plates set at each side of a stage, yet with no source of light apparent.*

But the light show was merely a warm-up for the inventor's guests.

---

* To this day no one has duplicated this demonstration.

Lines of tension on Tesla's face betrayed the seriousness with which he himself regarded the next experiment.

A small animal was brought from a cage, tied to a platform, and quickly electrocuted. The indicator registered one thousand volts. The body was removed. Then Tesla, with one hand in his pocket, leaped lightly upon the same platform. The voltage indicator began slowly climbing. At last two million volts of electricity were pouring "through" the frame of the tall young man, who did not move a muscle. His silhouette was now sharply defined with a halo of electricity formed by myriad tongues of flame darting out from every part of his body.

Seeing the shock on McGovern's face, he extended one hand to the English interviewer, who described the strange sensation: "You twist it about in the same fashion as you have seen people do who hold the handles of a strong electric battery. The young man is literally a human electric 'live wire.'"

The inventor leaped down from the platform, turned off the current, and relaxed the tension of his audience by tossing off the performance as no more than a trick. "Pshaw! These are only a few playthings. None of these amount to anything. They are of no value to the great world of science. But come over here, and I will show you something that will make a big revolution in every hospital and home as soon as I am able to get the thing into working form."

He led his guests to the corner where a strange platform was mounted on rubber padding. When he flipped a switch, it began to vibrate rapidly and silently.

Twain stepped forward, eager. "Let me try it, Tesla. Please."

"No, no. It needs work."

"Please."

Tesla chuckled. "All right, Mark, but don't stay on too long. Come off when I give you the word." He called to an attendant to throw the switch.

Twain, in his usual white suit and black string tie, found himself humming and vibrating on the platform like a gigantic bumblebee. He was delighted. He whooped and waved his arms. The others watched in amusement.

After a time the inventor said, "All right, Mark. You've had enough. Come down now."

"Not by a jugful," said the humorist. "I am enjoying this."

"But seriously, you had better come down," insisted Tesla. "Believe me, it is best that you do so."

Twain only laughed. "You couldn't get me off this with a derrick."

The words were scarcely out of his mouth when his expression froze. He lurched stiffly toward the edge of the platform, frantically waving at Tesla to stop it.

"Quick, Tesla. Where is it?"

The inventor helped him down with a smile and propelled him in the direction of the rest room. The laxative effect of the vibrator was well known to him and his assistants.[4]

None of his guests had volunteered to undergo the experiment in which Tesla stood on the high-voltage platform; they never did. But now they clamored for an explanation of why he had not been electrocuted.

As long as the frequencies were high, he said, alternating currents of great voltages flowed largely on the outer surface of the skin without injury. But it was no stunt for amateurs, he warned. Milliamperes penetrating nerve tissue could be fatal, while amperes distributed over the skin could be tolerated for short periods. Very low currents flowing beneath the skin, whether alternating current or direct current, could kill.

It was dawn when Tesla finally said good-night to his guests. But the lights burned on in his laboratory for another hour before he locked the doors and walked to his hotel for a brief period of rest.

2. A GAMBLING MAN

Nikola Tesla was born at precisely midnight between July 9 and 10, 1856, in the village of Smiljan, province of Lika, Croatia, between Yugoslavia's Velebit Mountains and the eastern shore of the Adriatic Sea. The tiny house in which he was born stood next to the Serbian Orthodox Church presided over by his father, the Reverend Milutin Tesla, who sometimes wrote articles under the nom-de-plume "Man of Justice."

No country in Eastern Europe had greater ethnic and religious diversity than Yugoslavia. Within Croatia the Serbian Teslas were part of a racial and religious minority. The province then belonged to the Austro-Hungarian Empire of the Hapsburgs to whose heavy-handed rule the people adapted as best they could.

Ethnic traditions are often most tenaciously observed by transplanted minorities and the Teslas were no exception. They placed great store on Serbian martial songs, poetry, dancing, and storytelling, as well as on weaving and the celebration of saints' days.

Although illiteracy was more common than not in that time and place, it was of a rare mind-expanding kind, for the people both admired and cultivated prodigious feats of memory.

In the Croatia of Tesla's childhood, choices of career were more or less limited to farming, the Army, or the Church. The families of Milutin Tesla and his wife Đuka Mandić, who came originally from western Serbia, had for generations sent their sons to serve Church or Army and their daughters to marry ministers or officers.

Milutin had originally been sent to Army officers' school, but he had rebelled and left to join the ministry. This he saw as the only career for his sons, Dane (or Daniel) and Nikola. As for their sisters, Milka, Angelina, and Marica, the Reverend Tesla hoped that God in His wisdom and mercy would provide them with clerical husbands like himself.

The life of a Yugoslav woman was grueling, for she was expected not only to do the heavy work of the farm but also to raise the children and care for the home and family. Tesla always said that he inherited his photographic memory and his inventive genius from his mother, and deplored that she had not lived in a country and at a time when women's abilities were fairly rewarded. She had been the eldest daughter in a family of seven children, forced to take over when her mother became blind. Hence she herself got no schooling. But either in spite or because of that, she had developed an amazing memory, being able to recite verbatim whole volumes of native and classic European poetry.

After her marriage, her own five children arrived quickly. The eldest was Daniel. Nikola was the fourth.

Since the Rev. Milutin Tesla wrote poetry in his spare time, the boy grew up in a household where cadence always permeated ordinary speech and where the quoting of passages from the Bible or poetry was as natural as roasting corn over charcoal in summer.

In his youth Nikola also wrote poetry and would later take some to America with him. He would never permit his poems to be published, however, considering them too personal. When he grew older, it would delight him to astonish new friends by reciting their native poetry (in English, French, German, or Italian) at impromptu meetings. He continued to write an occasional poem throughout his life.

The child began when only a few years of age to make original inventions. When he was five, he built a small waterwheel quite unlike those he had seen in the countryside. It was smooth, without paddles, yet it spun evenly in the current. Years later he was to recall this fact when designing his unique bladeless turbine.

But some of his other experiments were less successful. Once he perched on the roof of the barn, clutching the family umbrella and hyperventilating on the fresh mountain breeze until his body felt light and the dizziness in his head convinced him he could fly. Plunging to earth, he lay unconscious and was carried off to bed by his mother.

His sixteen-bug-power motor was, likewise, not an unqualified success. This was a light contrivance made of splinters forming a windmill, with a spindle and pulley attached to live June bugs. When the glued insects beat their wings, as they did desperately, the bug-power engine prepared to take off. This line of research was forever abandoned however when a young friend dropped by who fancied the taste of June bugs. Noticing a jarful standing near, he began cramming them into his mouth. The youthful inventor threw up.

He next endeavored to take apart and reassemble the clocks of his grandfather. This too, he recalled, came to an end: "In the former operation I was always successful but often failed in the latter." Thirty years passed before he would tackle clockwork again.

Not all his youthful chagrins were scientific in nature. "There was a wealthy lady in town," he later recalled in a brief autobiography, "a good but pompous woman, who used to come to the church gorgeously painted up and attired with an enormous train and attendants. One Sunday I had just finished ringing the bell in the belfry and rushed downstairs when this grand dame was sweeping out and I jumped on her train. It tore off with a ripping noise which sounded like a salvo of musketry fired by raw recruits."[1]

His father, although livid with rage, gave him only a gentle slap on the cheek—"the only corporal punishment he ever administered to me but I almost feel it now." Tesla said his embarrassment and confusion were indescribable, and he was practically ostracized.

However, good fortune threw him a rope, and he was redeemed in the eyes of the village. A new fire engine had been purchased, along with uniforms for a fire department, and this called for a celebration. The community turned out for a parade, there were speeches, and then the command was given to pump water with the new equipment. Not a drop came from the nozzle. While the village fathers stood in puzzled dismay, the bright lad flung himself into the river and found, as he had suspected, that the hose had collapsed. He corrected the problem, instantly drenching the delighted village fathers. Long after, Tesla would recall that "Archimedes running naked thru the streets of Syracuse did not make a greater impression than myself. I was carried on the shoulders and was the hero of the day."[2]

In bucolic Smiljan where his first few years were spent, the intense child with the pale wedge-shaped face and shock of black hair seemed to live a charmed life. Just as in later years he would work with high voltages of electricity without serious harm, he then skated through extraordinary dangers.

With telescopic memory and perhaps some exaggeration, he later wrote that he was given up by doctors as a hopeless physical wreck three times, that he was almost drowned on numerous occasions, was nearly boiled alive in a vat of hot milk, just missed being cremated, and was once entombed (overnight in an old shrine). Hair-raising flights from mad dogs, enraged flocks of crows, and sharp-tusked hogs spiced this catalogue of near-catastrophes.[3]

Yet outwardly his parents' home provided an idyllic pastoral scene. Sheep grazed in the pasture, pigeons cooed in a cote, and there were chickens for a small boy to tend. Each morning he delighted in watching the flock of geese that rose magnificently to the clouds; they returned from the feeding grounds at sundown "in battle formation, so perfect that it would have put a squadron of the best aviators of the present day to shame."

For all of this outward beauty, however, there were ogres in the boy's mind, the lasting trauma of a family tragedy. As far back as he could remember, his life had been profoundly influenced by his older brother, who was seven at the time of Nikola's birth. Daniel, brilliant and the idol of his parents, was killed at the age of twelve in a mysterious accident.

The immediate cause of the tragedy may have been a magnificent Arabian horse which had been given to the family by a dear friend. It was petted by them and attributed with almost human intelligence. In fact this beautiful creature had once saved the father's life in the wolf-infested mountains. But according to Tesla's autobiography, Daniel died of injuries caused by the horse. Of the incident itself, however, no details remain.[4]

Anything Nikola did thereafter, he claimed, seemed dull by comparison to the promise of the dead brother. His own achievements "merely
caused my parents to feel their loss more keenly. So I grew up with little
confidence in myself. But I was far from being considered a stupid
boy. ..."

A second more psychologically intricate version exists as to how
Tesla's older brother died. According to the second version, Daniel died
from a fall down the cellar stairs. Some believe that the boy lost consciousness and in his delirium accused Nikola of pushing him. He died
later from the head injury, probably a hematoma, so this account goes.
Unfortunately at this date both versions are impossible to confirm.

Much later in his life, Tesla still suffered from nightmares and hallucinations related to the death of his brother. The details of the experience
are never clarified, but the episode recurs and is recounted throughout
his life as if from various time frames. One can theorize that a five-year-
old child, unable to tolerate such a burden of assumed guilt, might have
rewritten the facts in his mind.

We can only speculate about the degree to which Daniel's death may
have been responsible for the fantastic array of phobias and obsessions
that Nikola subsequently developed. All we can say for certain is that
some manifestations of his extreme eccentricity seem to have appeared at
an early age.

For example, he had a violent aversion to earrings on women,
especially pearls, although jewelry with the glitter of crystals or
sharp-planed facets intrigued him. The smell of a piece of camphor
anywhere in the house caused him acute discomfort. In research, if he
dropped little squares of paper in a dish filled with liquid, it caused a
peculiar and awful taste in his mouth. He counted steps when walking, calculated the cubic contents of soup plates, coffee cups, and
pieces of food. If he failed to do so his meal was unenjoyable—hence
his preference for dining alone. And perhaps most serious insofar as
physical relationships were to be concerned, he claimed that he could
not touch the hair of other people, "except perhaps at the point of a

revolver."[5] But we cannot precisely date the onset of these or his many other phobias.

According to Tesla, hoping to console his parents for the loss of Daniel, he subjected himself at a very early age to iron discipline in order to excel. He would be more spartan, more studious than other boys, more generous, and in every way superior. And it was while denying himself and repressing natural impulses, he later believed, that he began to develop his strange compulsions.

If Tesla's character did begin to change, the symptoms were not entirely apparent until some time after Daniel's death. "Up until the age of eight years," he wrote, "my character was weak and vacillating." He dreamed of ghosts and ogres, feared life, death, and God. But then there did come a kind of change, as the result of his favorite pastime—which was reading in his father's well-stocked library. The Rev. Milutin Tesla at one point forbade Nikola to have candles, fearing that he would ruin his eyes by reading all night. The boy got some materials and made his own, stuffed rags in the keyhole and door cracks, and then read all night. He did not stop reading until he heard his mother beginning her arduous rounds at dawn.

The book that changed his vacillating nature was *Abafi* or *The Son of Aba,* by a leading Hungarian novelist—a work that "somehow awakened my dormant powers of will and I began to practice self-control." To the rigorous discipline then developed, he attributed his later success as an inventor.[6]

From birth he was intended for the clergy. Although he longed to become an engineer, his father was inflexible. To prepare him, the Reverend Tesla initiated a daily routine: "It comprised all sorts of exercises—as guessing one another's thoughts, discovering the defects of some form or expression, repeating long sentences or performing mental calculations. These daily lessons were intended to strengthen memory and reason and especially to develop the critical sense, and were undoubtedly very beneficial."[7]

Of his mother he wrote that she was "an inventor of the first order and would, I believe, have achieved great things had she not been so remote from modern life and its multifold opportunities. She invented and constructed all kinds of tools and devices and wove the finest designs from thread which was spun by her. She even planted the seeds, raised the plants, and separated the fibers herself. She worked indefatigably, from break of day till late at night, and most of the wearing apparel and furnishings of the home was the product of her hands."[8]

The brilliant Daniel, before his untimely death, had been subject to strong flashes of light that interfered with his normal vision during moments of excitement. A similar phenomenon plagued Tesla during most of his life, beginning in childhood.

He described it years later as "a peculiar affliction due to the appearance of images, often accompanied by strong flashes of light, which marred the sight of real objects and interfered with my thought and action. They were pictures of things and scenes which I had really seen, never of those I imagined. When a word was spoken to me the image of the object it designated would present itself vividly to my vision and sometimes I was quite unable to distinguish whether what I saw was tangible or not. This caused me great discomfort and anxiety. None of the students of psychology or physiology whom I have consulted could ever explain satisfactorily these phenomena. . . ."[9]

He theorized that the images resulted from a reflex action from the brain upon the retina under great excitation. They were not hallucinations. In the stillness of night, the vivid picture of a funeral he had seen or some other disturbing scene would thrust itself *before* his eyes, so that even if he jabbed his hand through it, it would remain fixed in space.

"If my explanation is correct," he wrote, "it should be possible to project on a screen the image of any object one conceives and make it visible. Such an advance would revolutionize all human relations. I am convinced that this wonder can and will be accomplished in time to come; I may add that I have devoted much thought to the solution of the problem."[10]

Since Tesla's time parapsychologists have studied subjects who purportedly *can* project their mental images onto rolls of unexposed photographic film. The direct transmission of thought onto electronic printers also is the subject of recent research.

To free himself of the tormenting images and to obtain temporary relief, the young Tesla began to conjure up imaginary worlds. Every night he would start on make-believe journeys—see new places, cities, and countries, live there, meet people and make friends, and "however unbelievable, it is a fact that they were just as dear to me as those in actual life and not a bit less intense in their manifestations."[11]

This he did constantly until the age of seventeen, when his thoughts turned seriously to invention. Then, to his delight, he found that he could visualize with such facility that he needed no models, drawings, or experiments, but could picture them all as real in his mind.

He recommended this method as far more expeditious and efficient than the purely experimental. Anyone who carries out a construct, Tesla held, runs the risk of becoming bogged down in the details and defects of the apparatus and, as the designer goes on improving, tends to lose sight of the underlying principle of the design.

"My method is different," he wrote. "I do not rush into actual work. When I get an idea I start at once building it up in my imagination. I change the construction, make improvements and operate the device in my mind. It is absolutely immaterial to me whether I run my turbine in my thought or test it in my shop. *I even note if it is out of balance.*"[12]

Thus, he claimed he was able to perfect a conception without touching anything. Only when all the faults had been corrected in his brain did he put the device into concrete form.

"Invariably," he wrote, "my device works as I conceived that it should, and the experiment comes out exactly as I planned it. In twenty years there has not been a single exception. Why should it be otherwise? Engineering, electrical and mechanical, is positive in results. There is scarcely a subject that cannot be mathematically treated and the effects calculated or the results determined beforehand from the available theoretical and practical data. . . ."[13]

Despite such claims, Tesla did in fact often make small sketches of inventions in whole or in part. Later in life his methods of research came to resemble more closely the empirical approach of Edison.

Tesla's childhood development is confusing because he enhanced his native talent with such rigorous mental discipline that it is impossible to separate the innate gifts from the acquired. Some people, for example, prefer to think of Tesla's prodigious memory as being in no way abnormal but merely the result of making the most of what God gave him. Yet the ability to memorize a page of type or the precise relationships and sizes of myriad patterns on a page in the wink of an eye—call it photographic, eidetic, or whatever—does seem to belong to the specially gifted. Such memory usually begins to wane in adolescence, indicating that it is affected by bodily chemical changes.

In Tesla's case, perhaps because of his special training in early childhood and his subsequent self-discipline, his phenomenal memory was retained throughout much of his life. The fact that he began to make trial-and-error adjustments of his research equipment in Colorado when he was middle-aged hints at a waning power.

He claimed that his method of visual invention had one defect that kept him poor in a monetary sense, though rich in the raptures of the mind: Potentially valuable inventions were often put aside without the final time-consuming perfection required for commercial success. Edison would never have allowed this to happen and hired many assistants to make sure it did not. In fact Edison was said to have a knack for picking up other inventors' ideas and rushing them to the Patent Office. With Tesla it was to be just the opposite. Ideas chased each other through his mind faster than he could nail them down. Once he understood exactly how an invention worked (in his mind), he tended to lose interest, for there were always exciting new challenges just over the horizon.

His photographic memory explained in part the lifelong difficulty he would experience in working with other engineers. While they demanded blueprints, he worked in his mind. In grade school he was almost kept back, despite brilliance in mathematics, because he so loathed the required drawing classes.

He was twelve years old before he succeeded in banishing disturbing images from his mind by deliberate effort; but he was never able to control the inexplicable flashes of light that usually occurred when he was in a dangerous or distressing situation, or when he was greatly elated. Sometimes he saw all the air around him filled with tongues of living flame. Their intensity, instead of diminishing, increased with years and reached a peak when he was about twenty-five.

At sixty he reported, "These luminous phenomena still manifest themselves from time to time, as when a new idea opening up possibilities strikes me, but they are no longer exciting, being of relatively small intensity. When I close my eyes, I invariably observe first, a background of very dark and uniform blue, not unlike the sky on a clear but starless night. In a few seconds this field becomes animated with innumerable scintillating flakes of green, arranged in several layers and advancing towards me. Then there appears, to the right, a beautiful pattern of two systems of parallel and closely spaced lines, at right angles to one another, in all sorts of colors with yellow-green and gold predominating. Immediately thereafter the lines grow brighter and the whole is thickly sprinkled with dots of twinkling light. This picture moves slowly across the field of vision and in about ten seconds vanishes to the left, leaving behind a ground of rather unpleasant and inert grey which quickly gives way to a billowy sea of clouds, seemingly trying to mould themselves in living shapes. It is curious that I cannot project a form into this grey until the second phase is reached. Every time, before falling asleep, images of persons or objects flit before my view. When I see them I know that I am about to lose consciousness. If they are absent and refuse to come it means a sleepless night." [14]

In school he excelled at languages, learning English, French, German, and Italian as well as the Slavic dialects, but it was math at which he starred. He was that unnerving sort of student who lurks behind the instructor while problems are being written on the board, and quietly chalks down answers the moment the teacher has finished. At first they suspected him of cheating. But soon it was realized that this was just another aspect of his abnormal ability to visualize and retain images. The optic screen in his mind stored entire logarithmic tables to be called on as needed. After he became an inventor, however, he would sometimes have to struggle for long periods to solve a single scientific problem.

He reported another curious phenomenon that is familiar to many creative people, i.e., that there always came a moment when he was not concentrating but when he *knew* he had the answer, even though it had not yet materialized. "And the wonderful thing is," he said, "that if I do feel this way, *then I know I have really solved the problem and shall get what I am after.*"

Practical results generally confirmed this intuition. It is a fact that in later life the machines that Tesla built nearly always worked. He might err in his understanding of the scientific principle, or he might even mistake the quality of materials used in construction, but somehow the machines, as they evolved in his mind and were later translated into metal, usually did just what he intended.

Had there been school psychologists in his childhood, the bedeviling images that warred with his sense of reality might easily have earned him a diagnosis of schizophrenia, and therapy or drugs might have been prescribed—perhaps to "cure" the very fountain of his creativity.

When he first discovered that the pictures in his mind could always be traced to actual scenes which he had previously observed, he believed he had hit upon a truth of great importance. He made a point of always trying to trace the external source. In short, before Freud's methods were well known, he was practicing a kind of autoanalysis, and after a while this effort grew to be almost reflexive.

"I gained great facility in connecting cause and effect," he reported. "Soon I became aware, to my surprise, that every thought I conceived was suggested by an external impression." [15]

The conclusion he drew from this exercise was not altogether heartening. Everything he did that he had thought to be the result of free will he now decided was actually caused by real circumstances and events. And if this were true it followed that he himself must be merely a kind of automaton. Conversely, anything a human being could do, a machine could be made to do, including acting with judgment based upon experience.

From these meditations the young Tesla developed two concepts that—in rather different ways—were to be important to him in later life. The first was that human beings could be adequately understood as "meat machines." The second was that machines could, for all practical purposes, be made human. The first idea may have done nothing to improve his sociability, but the second was to lead him deep into the strange world of what he called "teleautomatics" or robotry.

The Teslas had moved to the nearby city of Gospić when Nikola was six. There he entered school and had seen his first mechanical models, including water turbines. He built many of them and found great pleasure in operating them. He also became fascinated by a description he had read of Niagara Falls. In his imagination a big wheel appeared, run by the cascading waters. He told his uncle that one day he would go to America and carry out this vision. Thirty years later, seeing his idea materialize, Tesla would marvel "at the unfathomable mystery of the mind."

At ten he entered the gymnasium, which was new and had a fairly well-equipped department of physics. The demonstrations performed by his instructors fascinated him. Here his brilliance in mathematics shone, but his father "had considerable trouble in railroading me from one class to another" because he could not endure the required course in freehand drawing.

In the second year he became obsessed with the idea of producing continuous motion through steady air pressure, and with the possibilities of a vacuum. He grew frantic with his desire to harness these forces but for a long time groped in the dark. Finally, he recalled, "my endeavors crystallized in an invention which was to enable me to achieve what no other mortal ever attempted." It was all part of his consuming dream of being able to fly.

"Every day I used to transport myself through the air to distant regions but could not understand just how I managed to do it," he recalled. "Now I had something concrete—a flying machine with nothing more than a rotating shaft, flapping wings, and . . . a vacuum of unlimited power!"[16]

What he built was a cylinder freely rotatable on two bearings and partly surrounded by a rectangular trough which fit it perfectly. The open side of the trough was closed by a partition and the cylindrical segment divided into two compartments entirely separated from each other by airtight sliding joints. One of these compartments being sealed and exhausted of air, the other remaining open, perpetual rotation of the cylinder would result—or so the inventor thought. And indeed, when he had finished, the shaft rotated slightly.

"From that time on I made my daily aerial excursions in a vehicle of comfort and luxury as might have befitted King Solomon," he recalled. "It took years before I understood that the atmospheric pressure acted at right angles to the surface of the cylinder and that the slight rotary effort I observed was due to a leak. Though this knowledge came gradually it gave me a painful shock."[17]

While at this school—for which he was probably too advanced— he was prostrated "with a dangerous illness, or rather a score of them, and my condition became so desperate that I was given up by physicians." To aid his recovery after there had been a change for the better, he was allowed to read. Finally he was asked to catalog the books at the local library, a task which, he later recalled, introduced him to the earliest works of Mark Twain. To his delight in finding them he attributed a miraculous recovery. Unfortunately, the anecdote has the ring of apocrypha, for Twain at that time had written almost nothing that might have found its way across the ocean and into a Croatian library. Whatever the truth of the story, Tesla liked it and stuck by it. Twenty-five years later he met the great humorist in New York City, told him of the experience, and was amazed, he said, to see Twain burst into tears.

The boy continued his studies at a higher school in Karlstadt (Karlovac), Croatia, where the land was low and marshy and where, in consequence, he suffered repeated bouts with malaria. Yet his illnesses did not prevent him from conceiving an intense interest in electricity under the stimulating influence of his physics professor. Every experiment he saw produced "a thousand echoes" in his mind, and he longed for experimentation and investigation as a career.

When next he returned home, a cholera epidemic was raging, and he immediately contracted the disease. He was in bed for nine months, scarcely able to move, and for the second time it was thought he was dying. He remembered that his father sat by his bed, trying to cheer him, and that he rallied sufficiently to suggest, "Perhaps I may get well if you will let me study engineering." The Reverend Tesla, who had never once relented in his determination that Nikola should enter the clergy, was now trapped by his own compassion and yielded.

What happened next is a little unclear. Apparently Tesla was summoned to serve for three years in the army, a prospect even more repugnant to him than the clergy. But in later life he did not refer to this, saying only that his father insisted he spend a year camping and hiking in the mountains to regain his health. In the event, he did spend a year in the latter fashion and did not serve in the army. His father's family included high-ranking officers, and in all likelihood their influence was employed to formalize his release from conscription for medical reasons.[18]

His rugged year in the mountains did nothing to subdue his fertile imagination. One plan he conceived was to build a tube under the Atlantic Ocean through which to shoot mail between the continents. He worked out mathematical details of a pumping plant to force water through the tube, which would push the spherical containers of mail. But he failed to gauge accurately the frictional resistance of the pipe to the flow of water. It appeared to be so great that he was forced to abandon the plan. Even so, he gained knowledge from this that would be applied in a later invention.

Never one to waste time on trifling schemes, he then conceived of building a gargantuan elevated ring around the equator. At first it would have scaffolding. Once this was knocked away the ring would rotate freely at the same speed as the Earth. In this respect it would have had its analogs in the synchronized satellites not invented until the late twentieth century. Tesla's goal, however, was even more ambitious. He proposed next to employ some reactionary force that would make the ring hold still with relationship to the Earth. Thus travelers could climb aboard it and be sped around Earth at a dizzying speed of 1,000 miles per hour—or rather, Earth would race beneath them, enabling them to circle the globe in a day while sitting still.

At the end of this magnificent, if impractical, year of wandering and dreaming, he was enrolled in 1875 at the Austrian Polytechnic School in Graz. During his first year he had a fellowship from the Military Frontier Authority and hence had no financial worries. Nevertheless, he crammed from three in the morning until eleven at night, determined to complete two years' work in one. Physics, mathematics, and mechanics were his main studies.

He records that the compulsion to finish everything, once started, almost killed him when he began reading the works of Voltaire. To his dismay he learned that there were close to one hundred volumes in small print "which that monster had written while drinking seventy-two cups of black coffee per diem." But there could be no peace for Tesla until he had read them all.

At the end of the year he sailed through nine exams with ease. But when he returned the following year his comfortable financial situation had evaporated. The Military Frontier was being abolished, there would be no fellowship, and the salary of a clergyman would be unable to cover the high tuition costs. Tesla would thus be obliged to drop out before the school year ended. He made the most of the little time he

had, however, and it was in this second year that he first began to toy
with the idea of an alternative to direct-current electrical machines.

The man responsible for introducing Tesla to the fascinations of
electrical machinery was a German, one Professor Poeschl, who
taught theoretical and experimental physics. Although he had
"enormous feet and hands like the paws of a bear," Tesla found his
experiments inspiring. When one day there arrived from Paris a
direct-current apparatus called a Gramme Machine that could be used
both as a motor and a dynamo, Tesla examined the machine intently,
feeling a strange excitement. It had a wire-wound armature with a
commutator. While operating, it sparked badly, and Tesla brashly suggested to Professor Poeschl that the design might be improved by dis-
pensing with the commutator and by switching to alternating current.

"Mr. Tesla may accomplish great things," the German scholar
retorted heavily, "but he will never do this. It would be equivalent to
converting a steadily pulling force, like that of gravity, into a rotary
effort. It is a perpetual motion machine, an impossible idea."[19]

The young Serb had no idea how it might be done, but instinct told
him that the answer already lay somewhere in his mind. He knew that he
would be unable to rest until he had found the solution.

But now Tesla's money had run out. He tried in vain to borrow, and
when this failed, he began to gamble. Although he was not a very good
card player, he became almost professionally skillful at billiards.

Unfortunately, his newfound skills did not save him. Tesla's
nephew, Nikola Trbojevich, says he was told by other family members
that Tesla was "fired" from the college and from the city by the police
as well "because of playing cards and leading an irregular life." The
nephew adds: "His mother got the money together for him to go to
Prague, as his father would not speak to him. In Prague, where he
spent two years, he might have gone to the university unofficially, but
the search made by the Czechoslovak Government shows that he was
not enrolled in any one of the four universities in Czechoslovakia. . . .
[I]t appears that Tesla was substantially a self-taught man, which by

no means detracts from his stature. Faraday also was a self-taught man."[20]

In 1879 Tesla had tried to find a job in Maribor but was unsuccessful. He was finally forced to return home. His father died that same year, and shortly thereafter he returned to Prague in the hope of being able to continue his studies. It is believed that until the age of twenty-four he remained there, auditing courses and studying in the library and so keeping abreast of progress in electrical engineering and physics.

Probably he continued gambling in an effort to keep in funds, but by this time he was well free of any danger of becoming an addict. Tesla himself has described how he became a gambler and then managed to reform. "To sit down to a game of cards," he recalled, "was for me the quintessence of pleasure. My father led an exemplary life and could not excuse the senseless waste of time and money in which I indulged. . . . I would say to him, 'I can stop whenever I please but is it worthwhile to give up that which I would purchase with the joys of Paradise?' On frequent occasions he gave vent to his anger and contempt, but my mother was different. She understood the character of men and knew that one's salvation could only be brought about through his own efforts. One afternoon, I remember, when I had lost all my money and was craving for a game, she came to me with a roll of bills and said, 'Go and enjoy yourself. The sooner you lose all we possess the better it will be. I know that you will get over it.' She was right. I conquered my passion then and there. . . . I not only vanquished but tore it from my heart so as not to leave even a trace of desire. . . ."[21]

Later in life he began to smoke excessively, and also found that the consumption of coffee was affecting his heart. Willpower triumphed again, and he banished both vices. He even stopped drinking tea. Obviously Tesla distinguished between the exercise of free will (which human "meat machines" lacked), and of willpower or the exercise of determination.

3. IMMIGRANTS OF DISTINCTION

Telegraphs were in operation in the United States and Europe. The transatlantic cable had been laid. Alexander Graham Bell's telephone was sweeping the Continent when the news came in 1881 that an exchange would soon be opened at Budapest. It was one of four cities chosen to be so honored by Thomas Alva Edison's European subsidiary.

Tesla left for Budapest in January of that year. He at once found a job, with the help of an influential friend of his uncle's, in the Central Telegraph Office of the Hungarian government. It was certainly not what the young electrical engineer would have chosen, being a drafting position at very low pay. However, with his usual zest he threw himself into the work.

Then he was stricken by a bizarre affliction which, for lack of a better name, his doctors called a nervous breakdown.

Tesla's senses had always been abnormally acute. He claimed that several times in boyhood he had saved neighbors from fires in their own homes when he was awakened by the crackling of flames. When he was past forty and carrying on his lightning research in Colorado, he would claim to hear thunderclaps at a distance of 550 miles, although the limit for his young assistants was 150 miles.

But what happened during his breakdown was astonishing even by Tesla standards. He could hear the ticking of a watch from three rooms away. A fly lighting on a table in his room caused a dull thud in his ear. A carriage passing a few miles away seemed to shake his whole body. A train whistle twenty miles distant made the chair on which he sat vibrate so strongly that the pain became unbearable. The ground under his feet was constantly trembling. In order for him to rest, rubber cushions were placed under his bed.

"The roaring noises from near and far," he wrote, "often produced the effect of spoken words which would have frightened me had I not been able to resolve them into their accidental components. The sun's rays, when periodically intercepted, would cause blows of such force on my brain that they would stun me. I had to summon all my willpower to pass under a bridge or other structure as I experienced a crushing pressure on the skull. In the dark I had the sense of a bat and could detect the presence of an object at a distance of twelve feet by a peculiar creepy sensation on the forehead."[1]

During this period his pulse fluctuated wildly from subnormal to 260 beats per minute. The continuous twitching and trembling of his own flesh became, in itself, a nearly unbearable burden.

Understandably the medical profession of Budapest was fascinated. A renowned doctor prescribed large doses of potassium while at the same time pronouncing the ailment unique and incurable.

Tesla writes, "It is my eternal regret that I was not under the observation of experts in physiology and psychology at that time. I clung desperately to life, but never expected to recover."[2]

Yet not only did his health return but, with the assistance of a devoted friend, he soon recovered greater vigor than ever. The friend was Anital Szigety, a master mechanic with whom Tesla often worked and an athlete. Szigety convinced him of the importance of exercise and, during this period, the two often went for long walks through the city.

In the years since he had left the Polytechnic at Graz, Tesla had never ceased to struggle with the problem of the unsatisfactory direct-current machine. He later wrote, in his usual flamboyant way, that he did not undertake the problem with a simple resolve to succeed. "With me it was a sacred vow, a question of life and death. I knew that I would perish if I failed."

But in fact he already sensed that the battle was won. "Back in the deep recesses of the brain was the solution, but I could not yet give it outward expression."[3]

One afternoon toward sunset, he and Szigety were walking in the

city park, and Tesla was reciting Goethe's *Faust*. The sinking sun
reminded him of a glorious passage:

> The glow retreats, done in the day of toil;
> It yonder hastes, new fields of life exploring;
> Ah, that no wing can lift me from the soil,
> Upon its track to follow, follow soaring!

Then, "the idea came like a flash of lightning, and in an instant the
truth was revealed."

Tesla's long, waving arms froze in midair as if he had been seized with
a fit. Szigety, alarmed, tried to lead him to a bench, but Tesla would not sit
until he had found a stick. Then he began to draw a diagram in the dust.

"See my motor here; watch me reverse it," he exclaimed.

The diagram that he drew would be shown six years later in his
address before the American Institute of Electrical Engineers, introducing to the world a new scientific principle of stunning simplicity and utility. The applications of it would literally revolutionize the technical
world.

It was an entire new system that he had conceived, not just a new
motor, for Tesla had hit upon the principle of the rotating magnetic field
produced by two or more alternating currents out of step with each other.[4]
By creating, in effect, a magnetic whirlwind produced by the out-of-step
currents, he had eliminated both the need for a commutator (the device
used for reversing the direction of an electric current) and for brushes providing passage for the current. He had refuted Professor Poeschl.

Other scientists had been trying to invent AC motors but had used
only a single circuit, just as in direct current, which either would not work
or worked badly, churning up a great deal of useless vibration. Alternating
currents were being used to feed arc lights as early as 1878–79 by Elihu
Thomson, who built a generator in the United States. The Europeans,
Gaulard and Gibbs, had produced the first alternating-current transformer, which was necessary for increasing and decreasing voltages in

power transmission. George Westinghouse, an early advocate of AC with great plans for the electrification of America, bought the American rights to the Gaulard and Gibbs patents.

Yet with all this activity there had been no truly successful AC motor until Tesla invented his—an induction motor that was the heart of a new system and a quantum jump ahead of the times.

But of course it is one thing to create a significant invention and quite another to make people aware of it. Tesla had already begun to picture himself as rich and famous, a strong tribute to the power of imagination, since his paycheck barely sustained him. As he wryly observed, "the last twenty-nine days of the month were the hardest." But even hardship now seemed more tolerable for he knew that at last he could call himself an inventor.

"This was the one thing I wanted to be," he recalled. "Archimedes was my ideal. I admired the works of artists, but to my mind they were only shadows and semblances. The inventor, I thought, gives the world creations which are palpable, which live and work."[5]

In the days that followed he gave himself up entirely to the intense enjoyment of devising new forms of alternating-current machines.

"It was a mental state of happiness about as complete as I have ever known in life," he was to recall. "Ideas came in an uninterrupted stream, and the only difficulty I had was to hold them fast.

"The pieces of apparatus I conceived were to me absolutely real and tangible in every detail, even to the minutest marks and signs of wear. I delighted in imagining the motors constantly running. . . . When natural inclination develops into a passionate desire, one advances toward his goal in seven-league boots. In less than two months I evolved virtually all the types of motors and modifications of the system. . . ."[6]

He conceived of such practical alternating-current motors as polyphase induction, split-phase induction, and polyphase synchronous, as well as the whole polyphase and single-phase motor system for generating, transmitting, and utilizing electric current. And indeed, practically all electricity in the world in time would be generated, transmitted, distributed, and turned into mechanical power by means of the Tesla Polyphase System.

What it signified was vastly higher voltages than could be obtained through direct current and—with transmission possible over hundreds of miles—a new age of electric light and power everywhere. Edison's carbon filament light bulb could burn either AC or DC, but electricity couldn't be carried economically when a generator was required every two miles. And Edison was less adaptable than his light bulb, being emotionally locked into DC.

The year was 1882, and Tesla's ideas were still raging inside his head. Having neither the time nor the money for building prototypes, he turned his thoughts to the work of the telegraph office, where he was soon promoted to engineering. He made several improvements to the central-station apparatus (including inventing a telephone amplifier which he forgot to patent) and in return, the job gave him valuable practical experience.

Through family friends—two brothers named Puskas—he was next recommended for a job with Edison's telephone subsidiary in Paris, where he went in the fall of 1882.

Of paramount interest to him was to sell the officers of the Continental Edison Company on the enormous potential benefits of alternating current. The young Serb was bitterly disappointed, however, on being told of Edison's aversion to so much as the mention of this subject.

To be young and in Paris simultaneously provided opportunities for consolation that he did not overlook. He made new friends, both French and American, resumed his old proficiency at billiards, walked miles every day, and swam in the Seine.

At work he was given the job of troubleshooter, to cure the ills of Edison power plants in France and Germany. Sent to Alsace on a job for the firm, he took along materials, and there built his first actual alternating-current induction motor—"a crude apparatus, but [it] afforded me the supreme satisfaction of seeing for the first time, rotation effected by alternating currents without commutator."[7]

Twice during the summer of 1883 he repeated his experiments with the aid of an assistant. The advantages of AC over Edison's DC were so obvious to him that he could not believe anyone could close his eyes to them.

In Strassburg, Tesla was asked to see what could be done with a railroad-station lighting plant that the client, the German government, had refused to accept. And for good reason. A large chunk of wall had been blown out by a short circuit during the opening ceremony—in the presence of old Emperor William I. The French subsidiary, being faced with a serious financial loss, promised Tesla a bonus if he could improve the dynamos and soothe the Germans.

It was a ticklish operation for a relatively inexperienced person, but at least Tesla's ability to speak German helped. And in the end, not only was he able to correct the electrical problems, but he made friends with the mayor, one M. Bauzin, whom he then tried to recruit to support his invention. The mayor did in fact round up several wealthy potential investors to whom Tesla demonstrated his new motor. But although it functioned perfectly, the burghers simply could not see its practical advantages.

The disappointed young inventor was only partly consoled when the mayor produced some bottles of St. Estèphe 1801, left over from the last invasion of Alsace by the Germans. No one, he said kindly, was more worthy of the precious beverage than Tesla.

Having successfully completed his job, the inventor returned to Paris, looking forward to collecting his bonus. But to his dismay, it did not materialize. Of three administrators who were his superiors, each passed the buck to the next until Tesla, angered at being cheated, summarily resigned.[8]

The manager of the plant, Charles Batchelor, who had been a close friend and assistant of Edison's for many years, recognized the young Serb's abilities. He urged him to go to America where both grass and currency were greener.

Batchelor was an English engineer who had worked with Edison when the latter was improving Bell's first telephone. Edison had invented the transmitter that made it possible for voices to be heard over long distances, and it was Batchelor who helped him test the telephone in a boisterous public demonstration, uttering what a New York journalist described as "vociferous remarks and thunderous songs."

Subsequently, the Englishman and Edison together had supervised the installation of Edison's first commercial self-contained lighting plant on the S.S. *Columbia*, and the ship had made a brilliant display as she sailed down Delaware Bay on her voyage around Cape Horn to California.

Thus Batchelor had reason to think he knew Edison well, and he wrote Tesla a glowing letter of recommendation, introducing the one egocentric genius to the other. As events would prove, however, Batchelor understood Edison less well than he supposed.

"I liquefied my modest assets," Tesla later recalled, "secured accommodations and found myself at the railroad station as the train was pulling out. At that moment I discovered that my money and tickets were gone. What to do? Hercules had plenty of time to deliberate, but I had to decide while running alongside the train with opposite feelings surging in my brain like condenser oscillations. Resolve, helped by dexterity, won out in the nick of time...."[9]

He found enough change for the train and swung aboard. Later he talked his way aboard the ship *Saturnia* when no one showed up to claim his berth.

To America, beside the few coins in his pocket, he brought some poems and articles he had written, a package of calculations relating to what he described (without further elucidation) as an insoluble problem, and drawings for a flying machine. To be sure, at twenty-eight he was already one of the world's great inventors. But not another soul knew it.

4. AT THE COURT OF MR. EDISON

At least no one mistook Tesla in his smart bowler hat and black cutaway coat for a Montenegrin shepherd or fugitive from debtors' prison that June day he strode ashore at the Castle Garden Immigration Office in Manhattan. It was 1884, the year the people of France gave America the Statue of Liberty. As if in response to the words of Emma Lazarus, 16 million Europeans and Asians were to sweep into this country in a very few years, and they would keep coming. Men, women, and even children were needed as fuel to run America's fulminating industrial revolution. It was also the year of the Panic of 1884.

Tesla did not go to the Immigration hiring hall, where new arrivals were signed up for labor gangs to slave thirteen-hour days on the railroads, in mines, factories, or stockyards. Instead, with his letter of introduction to Edison and the address of an acquaintance in his pocket, he asked directions of a policeman and set out boldly onto the streets of New York.

Passing a shop where the owner was cursing at a broken machine, he stopped and offered to fix it. When he had done so, the man was so pleased that he gave Tesla twenty dollars.

As he walked on, the young Serb smiled to himself, remembering the joke that he had heard on shipboard. A Montenegrin shepherd who had just arrived in America was walking down the street when he saw a ten-dollar bill. He bent down to pick it up and then stopped, saying to himself, "My first day in America! Why should I work?"

Thomas Alva Edison, already graying at age thirty-two, buttoned to the chin in one of Mrs. Edison's hand-sewn, hand-styled, gingham smocks, was an ungainly, swinging, stooping, shuffling figure. At first glance his plain face might have seemed unremarkable, but it never took visitors long to be impressed by the light of fierce intelligence and relentless energy that shone in his eyes.

At the time, Edison was spread uncomfortably thin, even for a genius. He had opened the Edison Machine Works on Goerck Street and the Edison Electric Light Company at 65 Fifth Avenue. His generating station at 255–57 Pearl Street was serving the whole Wall Street and East River area. And he had a big research laboratory at Menlo Park, New Jersey, where a large number of men were employed and where the most astonishing things could happen.

Sometimes Edison himself could be seen there, dancing around "a little iron monster of a locomotive" that got its direct current from a generating station behind the laboratory, and which had once flown off the rails at a speed of forty miles per hour to the delight of its creator.[1] To this laboratory, also, Sarah Bernhardt had come to have her voice immortalized on Edison's phonograph. She had politely remarked upon his resemblance to Napoleon I.

The Pearl Street generating station served a few hundred individual mansions of wealthy New Yorkers with electric lights, but Edison also supplied direct current to isolated plants in mills, factories, and theaters all over the city. Also he was getting more and more requests to put lighting plants on ships—a particular headache since the danger of a fire at sea was a persistent nightmare.

And in addition to everything else, he still had to uphold his famous reputation as a man of pithy sayings: "Everybody steals in commerce and industry," went one of his apothegms. "I've stolen a lot myself. But I know *how* to steal. They don't know *how* to steal. . . ." *They* were Western Union, for whom he had worked while at the same time selling a competitive invention to their opposition.

There was also his contemptuous saying that he didn't need to be a mathematician because he could always hire them. Formally trained scientists might take umbrage, but at this particular stage of America's technological development there was no gainsaying that engineers and inventors probably *were* making more significant contributions to national life than their academic contemporaries. And just so no one would miss the point, Edison liked to add that he could always tell the importance of one of his inventions by the number of dollars it brought and that nothing else concerned him.

Julian Hawthorne observed, "If Mr. Edison would quit inventing and go in for fiction, he would make one of the greatest novelists. . . ."

On a particularly trying summer day in 1884, the American inventor had rushed straight from an electrical emergency at the Vanderbilt house on Fifth Avenue to his Pearl Street generating station. The house had caught fire from two wires that got crossed behind wall hangings that contained fine metallic thread. The flames had been smothered but Mrs. Vanderbilt, hysterical from the ordeal, had learned that the source of her problems was a steam engine and boiler in the cellar. Now the unreasonable woman was demanding that Edison remove the whole installation.[2]

He dispatched a repair crew, sucked a gulp of cold coffee from a mug, and tried to think what to do next. The telephone rang. Edison tilted the receiver to his good ear.

The manager of the shipping company that owned the S.S. *Oregon* sarcastically demanded to know if he had any plans for getting the dynamos repaired for his lighting plant. The liner had been tied up for days past sailing time and was losing bundles of money.

What could Edison say? He had no engineer to send.

He thought enviously of Morgan. Mr. J. Pierpont Morgan employed a full-time engineer just to run the private boiler and steam engine that was set into a pit below the garden of his Murray Hill mansion. It was so noisy the neighbors were threatening to sue. But that didn't bother Morgan; when things got too sticky, he could simply pack a supply of his favorite black cigars and set off for a nice long cruise on his yacht, the *Corsair.*

"I'll send an engineer over this afternoon," Edison promised the shipping magnate.

Morgan was the major financial backer of the Edison Electric Company, whose direct-current wires were festooned in localized, horse-frightening, malfunctioning webs above the streets of New York.

Although electricity was still little understood by the average financier or industrialist, a few like Morgan could see that it was easily the most promising development to have come along since Archimedes invented the screw. Everyone needed energy. And soon everyone would want Edison's incandescent lights.

Electrical engineering was *the* field for a gifted person of scientific or inventive bent to enter, offering not only financial reward but the seductiveness and danger of an almost unexplored frontier.

Cornell University and Columbia College were among the few schools in the country to boast fledgling departments of electrical engineering. America had only a handful of homegrown experts beyond such giants as Edison, Joseph Henry, and Elihu Thomson. Industrialists therefore would be glad to draw upon the foreign talent pool: Tesla, Michael Pupin, Charles Proteus Steinmetz, Batchelor, and Fritz Lowenstein, among others.

Yet it was primarily thanks to Edison's rough-and-ready ingenuity that the lights were flickering on (and off) in New York City. Only the year before, Mrs. William K. Vanderbilt had staged the epic ball that signaled peace at last between the feuding Astors and the Vanderbilts, and Mrs. Cornelius Vanderbilt had sailed down the grand staircase of the family mansion dressed as "The Electric Light," an apparition in white satin and diamonds that few at the ball would ever forget.

So glamorous was the new energy source that a manufacturer advertised at Christmas urging fathers to "Surprise the whole family with a double socket." Equally exciting—if puzzling—gifts guaranteeing to put one ahead of the Joneses were an electric corset for Mom and a magnetic belt for Dad. Yokels at county fairs were paying for the joy of getting a shock from a storage battery.

Edison had no sooner promised his nonexistent engineer to the shipping company and cradled the telephone receiver that June day than a breathless boy dashed into the shop to report trouble at Ann and Nassau streets. A junction box that had been wired by one of the inventor's inexperienced electricians was leaking. The boy vividly described how a ragman and his horse had been catapulted into the air and then had disappeared down the street at an unbelievable clip.

Edison bellowed for his foreman: "Get a gang of men, if you can find any. Cut off the current and fix that leak."

He glanced up and became aware of a tall dark presence hovering just inside his office.

"Help you, mister?"

Tesla introduced himself, speaking in careful accented English and a little louder than usual, for he knew of Edison's hearing problem.

"I have this letter from Mr. Batchelor, sir."

"Batchelor, eh? What's wrong in Paris?"

"Nothing that I know of, sir."

"Nonsense, there's always something wrong in Paris."

Edison read Batchelor's brief note of recommendation and snorted. But he gave Tesla a penetrating look.

"'I know two great men and you are one of them; the other is this young man!' Hmph! That's some recommendation. What can you do?"[3]

Tesla had rehearsed this moment many times on shipboard. Edison's reputation impressed him deeply. Here was a man who, without formal education of any sort, had invented hundreds of useful products. He himself had spent years digging away at books, but for what? What had he to show for it? What use was all his education?[4]

Quickly he began to describe the work he had done for Continental Edison in France and Germany. And then, before Edison could even respond, he moved smoothly into a description of his marvelous induction motor for alternating current, based upon his discovery of the rotating magnetic field. This was the wave of the future, he said. A smart developer could make a thousand fortunes with it.

"Hold up!" said Edison angrily. "Spare me that nonsense. It's dangerous. We're set up for direct current in America. People like it, and it's all I'll ever fool with. But maybe I could give you a job. Can you fix a ship's lighting plant?"

Tesla boarded the S.S. *Oregon* that same day with his instruments and began to make the necessary repairs. The dynamos were in bad condition, having several short circuits and breaks. With the aid of the crew he worked through the night. At dawn the next morning the job was finished.

As he walked back along Fifth Avenue toward the Edison shop, he met his new employer and a few of his top men just going home to rest.

"Here is our 'Parisian' running around at night," commented Edison.[5]

When Tesla said that he had just finished repairing both machines, Edison looked at him in silence, then walked away without another word. But the Serb with his acute hearing heard him remark at a little distance, "That is a damn good man."

Edison later told him about another important European scientist's arrival in the United States. Charles Proteus Steinmetz, the brilliant German dwarf, was almost deported as an indigent alien. He somehow squeaked through and went on to become the resident genius of General Electric's first industrial research laboratory at Schenectady. He would later strive to develop an acceptable alternative to Tesla's alternating-current system when Edison and General Electric needed to play catch-up.

Tesla's skills were quickly appreciated by Edison, who gave him almost complete freedom in working on the design and operating problems of the shop. He regularly worked from 10:30 in the morning until 5:00 the following morning, a regimen that won from his new boss the grudging comment, "I have had many hardworking assistants but you take the cake."

Both men had the ability in an emergency to go without sleep for two or three days while ordinary mortals crumpled around them. Edison's workers always claimed, however, that he sneaked catnaps.

Before long Tesla observed ways in which the primitive Edison dynamos could be made to work more efficiently, even though limited to the production of direct current. He proposed a plan for redesigning them and said it would not only improve their service but would save a lot of money.

The astute businessman in Edison brightened at the mention of the latter, but he realized the project Tesla had described was major and would take a long time. "There's fifty thousand dollars in it for you—if you can do it," he said.[6]

For months Tesla worked frenziedly, scarcely sleeping from one day to the next. In addition to redesigning the twenty-four dynamos completely and making major improvements to them, he installed automatic controls, using an original concept for which patents were obtained.

The personality differences between the two men doomed their relationship from the start. Edison disliked Tesla for being an egghead, a theoretician, and cultured. Ninety-nine percent of genius, according to the Wizard of Menlo Park, was "knowing the things that would not work." Hence he himself approached each problem with an elaborate process of elimination.

Of these "empirical dragnets" Tesla later would say amusedly, "If Edison had a needle to find in a haystack, he would proceed at once with the diligence of the bee to examine straw after straw until he found the object of his search. I was a sorry witness of such doings, knowing that a little theory and calculation would have saved him ninety percent of his labor."[7]

The well-known editor and engineer Thomas Commerford Martin recorded that Edison, unable to find Tesla's obscure birthplace in Croatia on a map, once seriously asked him whether he had ever eaten human flesh.

"Even the most cometic genius has its orbit," Martin wisely wrote, "and these two men are singularly representative of different kinds of training, different methods, and different strains. Mr. Tesla must needs draw apart . . . for his own work's sake."

In so basic a matter as personal hygiene they could not have been more different: Tesla, afraid of germs, fastidious in the extreme, once observed of Edison, "He had no hobby, cared for no sport or amusement of any kind and lived in utter disregard of the most elementary rules of hygiene. . . . [I]f he had not married later a woman of exceptional intelligence, who made it the one object of her life to preserve him, he would have died many years ago from consequences of sheer neglect. . . ."⁸

The irreconcilable differences, however, went beyond personality. Edison sensed the talented foreigner's threat to his direct-current system, erroneously thinking DC was vital to the manufacture and sale of his incandescent light bulbs. It was the old story of vested interest. At the beginning Edison himself had met with violent resistance from the gas monopolies. He had beaten down the gas companies with his natural gift for propaganda, putting out regular bulletins in which he gleefully described the dangers of gas-main explosions. His salesmen were sent out to cover the country, reporting every incident of "industrial oppression" in which workers' health allegedly had been "injured" by gas heat or their vision damaged by gaslights. Now it looked as if he might have to lash out against an even newer technology than his own.⁹

Tesla, in the odd moments of spare time he could grasp, was absorbing the history, literature, and customs of America, relishing new friendships and experiences. He already spoke English well and was even beginning to understand the American sense of humor. Or at least he thought he did. As events would prove, Edison still had a few things to teach him about that.

He enjoyed walking the streets of New York where the new, electrically powered trolleys brought congestion and not a little excitement to already jammed thoroughfares. Half the time the central dynamos were broken down. When the trolleys ran, they scared pedestrians as much as the passengers. The editor of a newspaper solemnly warned that anyone who rode on them might expect to be stricken with palsy and should look for no sympathy.

Brooklynites, who for some reason felt especially singled out for attack by vicious trolleys, banded together under the slogan of "Trolley Dodgers." Later, when the borough acquired a baseball team of its own, it seemed natural to call them the Brooklyn Dodgers.

It took Tesla the better part of the year to finish redesigning Edison's dynamos. When at last the job was done, he went to his boss to report complete success and, not incidentally, to ask when he might receive his $50,000.

Edison swept his high black shoes from his desk and fell forward openmouthed.

"Tesla," he exclaimed, "you don't understand our American humor."[10]

Once again it seemed that the Serb was to be deliberately cheated by an Edison company. Angered, he announced he would resign. Edison offered a compromise: a $10 raise of his princely salary of $18 per week. Tesla picked up his bowler hat and walked out.*

In Edison's view Tesla was "the poet of science"—his ideas "magnificent but utterly impractical." He warned the young engineer that he was making a mistake—and so it appeared for a time. The country was still deep in the gloom of financial crises with jobs hard to find.

Edison, completely in Morgan's grip, was himself having frustrating financial problems. While the inventor ached for full-speed-ahead, the banker insisted on a go-slow policy. He denied Edison even the most modest loans for expansion while the House of Morgan's capital was being poured into gigantic railroad acquisitions.

The process of "Morganization" had become standardized. Of everything he touched, the financier soon controlled 51 percent, and he insisted on being on the board of directors, however anonymously. Morganization meant the steady acquiring of companies engaged in a similar line of business, the sale of watered stock, and the centralizing of power through the elimination of "destructive competition."

Morgan, in his forties and near the peak of his power, was truculent, arrogant, feared, a loner who cared nothing for his associates, his underlings, or the public. He was six feet tall, weighed two hundred pounds, and because of an unfortunate skin disease, his nose glowed like one of Edison's newfangled light bulbs. Still, such is the power of power, he was a Don Juan whose conquests were openly flaunted.[11]

---

* The Edison camp has a different version: that Tesla offered to sell his AC patents to Edison for $50,000, and the latter jokingly declined.

His veneer of culture required frequent art-collecting trips to Europe, where he was more discriminating than many parvenus who amassed the treasures of the Old World. A staunch supporter of the Episcopal Church, he often left his Wall Street offices in the afternoon to spend a happy hour booming familiar hymns to the rafters at St. George's Episcopal Church, accompanied by his favorite organist.

Plagued by such evils as railroad rate wars and labor riots that threatened his rolling stock, he welcomed opportunities to escape from his desk. When traveling in America he rode in a $100,000 "palace car" attached to the train of his choice. More humble wheels were shunted from his path.

Like Edison he was noted for his pithy sayings. One that Tesla would have reason to recall was: "A man always has two reasons for the things he does—a good one and the real one."

The financial panic of '84 had caused such insecurity that thousands of small investors all over America were going broke. Businessmen turned to the powerful House of Morgan, rather than to government, for salvation. It looked to the financier as if all his careful plans for the centralizing of control over the economic machine might be wrecked by labor troubles and the rate wars among the overly expanded railroads.

It was clear to anyone that far too many railroads had been built for speculative purposes and that many were facing bankruptcy. There would have to be a merger. But Morgan was not a man to be pushed or to act rashly. Let his competitors sweat. He would visit the spas of Europe and collect art.

By midsummer of the year Tesla arrived in America, Morgan's leisurely travels had brought him to England, there to receive still more unpleasant reports from home of "railroad wrecking" and panic. Finally he consented to return and put his formidable brain to work for the sake of the Nation.

Morgan's solution was simply to summon all the quarreling capos to a peace conference aboard the *Corsair.*[12] All of one day he and the captive industrial barons cruised up and down the bay and the East River. This was no war of individuals but of competing oil, steel, and railroad inter-

ests locked in oligarchic struggle. Before the night fell, Morgan had
"reorganized" them all in such a truly masterful way that through clever
mergers he had reduced "destructive competition" to a minimum. This
was the essence of the Morgan touch, a touch that would soon make
itself felt in the promising new field of electrical utilities.

Meanwhile, Tesla, whose engineering reputation was beginning to
be favorably known, was approached by a group of investors and offered
a chance to form a company under his own name. He leaped at it. At last
his great alternating-current discovery could be presented to the world.
Humanity, as he saw it, would be freed from its burdens. Unfortunately,
his backers had something more modest and practical in mind. There was
a big market for improved arc lights for streets and factories, and this
would have to come first.

The Tesla Electric Light Company was formed, with headquarters at
Rahway, New Jersey, and a branch office in New York. One of the men
involved in this firm was James D. Carmen, who was to be a behind-the-
scenes ally of Tesla's for twenty years or more. He and Joseph H.
Hoadley would serve as officers in several of Tesla's companies.

Working in his first laboratory on Grand Street, the Serb developed
a Tesla arc lamp which was more simple, reliable, safe, and economical
than those in current use.[13] The system was patented and first put to
work on the streets of Rahway.*

Tesla's compensation was to have been shares of stock in the firm.
Now, to his painful surprise at the ways of American commerce, he
found himself being eased out of the company. He wound up with a
handsomely engraved stock certificate which, because of the newness of
the firm and the recurring economic crises, had little redeemable value.

Exit Tesla for the third time.

The slump became a depression, and he was unable to find an
engineering position. From the spring of 1886 until the following year
he went through one of the more depressing periods of his life.

* Patents 334,823, 335,786, 335,787, 336,961, 336,962, 359,954, 359,748.

Toiling as a laborer on New York street gangs, he barely managed to survive. Tesla seldom referred to this painful experience afterward.

Nevertheless he had made some progress: his arc-lighting innovations resulted in the granting of seven patents, and in addition he obtained other light-related patents, two of which are particularly interesting.* They involve using the loss of magnetism in iron at temperatures above 750 degrees Celsius, for transforming heat directly into mechanical or electrical energy. Like a number of Tesla's inventions, they found no immediate use and were forgotten. But quite recently in the twentieth century a similar process has gained attention, without recognition being given to Tesla's prior inventions.

Four years had passed since he had discovered the rotating magnetic field and constructed his first alternating-current motor at Strassburg. He was beginning to wonder whether the green pastures and golden promise of America would continue to elude him. Humiliated by recent disappointments, he again brooded upon what seemed like his wasted years of education.

But then his luck took another unexpected turn. Having heard of his induction motor, the foreman of the work crew on which the inventor was suffering so bitterly took him to meet A. K. Brown, manager of the Western Union Telegraph Company, who not only knew about alternating current but was personally interested in the new idea.

Where Edison had failed to see the revolution ahead or, more likely, had seen in it the death knell of his own direct-current system of electrification, Brown correctly gauged the future. With his help another company was created in Tesla's name. The Tesla Electric Company had the specific goal of at last developing the alternating-current system that the inventor had conceived in the park in Budapest in 1882.[14]

---

* Patents 396,121, Thermomagnetic Motor, and 428,057, Pyromagneto-Electric Generator. See also 382,845 Commutator for Dynamo-Electric Machine.

5. THE WAR OF THE CURRENTS
BEGINS

The laboratory and shops that the ecstatic Tesla found for his new company were at 33–35 South Fifth Street, only blocks from the Edison workshops. The Tesla Electric Company, capitalized with half a million dollars, opened for business in April 1887. To the inventor, who had waited so long for this moment, it was the fulfillment of a dream. He began laboring like one of his own dynamos, day and night without rest.

Because it was all there in his mind he needed only a few months to start filing patent applications for the entire polyphase AC system. This was in fact three complete systems for single-phase, two-phase, and three-phase alternating currents. He experimented with other kinds too. And for each type he produced the necessary dynamos, motors, transformers, and automatic controls.

Hundreds of central stations were operating in America at this time, using at least twenty different combinations of circuits and equipment. Usually these were centered upon one invention or group of them. Thus Elihu Thomson had installed a small alternator and transformers in the factory of the Thomson-Houston Company at Lynn, Massachusetts, in 1886, supplying incandescent lamps in another factory. But it was to be another year before he evolved a safe system for wiring houses. So, too, George Westinghouse, inventor of the railroad air brake, having acquired patents to the AC distribution system of Gaulard and Gibbs, set his chief engineer, William Stanley, to building a transformer system. It was successfully tested in 1886. Westinghouse operated the first commercial AC system in America at Buffalo in November of that year and by 1887 had more than thirty plants in operation. In addition there was of course the direct-current system of the Edison Electric Company, one of the earliest contenders in the field.

But still no satisfactory alternating-current motor existed. Within six months after opening his shop, Tesla sent two motors to the Patent Office for testing and filed his first AC patents.* In all, through 1891, he applied for and was granted a total of forty patents.† So original and sweeping were they that he met with no delay.[1]

And now, recognition was mercifully swift in coming. William A. Anthony, who had established a course in electrical engineering at Cornell University, saw the significance of the Tesla system at once and spoke out in its favor. This was not just a new motor but quite possibly the foundation of a new technology. The essence of the system, as Anthony noted, was the beautifully simple induction motor, which had almost no wearing parts to break down.

The news of such unheralded activity in the U.S. Patent Office rocked Wall Street as well as the industrial and academic worlds. At Professor Anthony's suggestion the almost unknown young Serb was invited to lecture to the American Institute of Electrical Engineers on May 16, 1888.

Tesla, to his surprise, discovered himself to be a natural and brilliant lecturer; and his address became a classic. His subject was "A New System of Alternate Current Motors and Transformers."[2]

Dr. B. A. Behrend, commenting on the presentation, said, "Not since the appearance of Faraday's 'Experimental Researches in Electricity' has a great experimental truth been voiced so simply and so clearly. . . . He left nothing to be done by those who followed him. His paper contained the skeleton even of the mathematical theory."[3]

Tesla's timing could not have been better. His patents were the missing key that George Westinghouse had been waiting for. The Pittsburgh magnate, a stocky, blunt, dynamic fellow with a walrus mustache, had a taste for fashionable dress and for adventure. Like

---

* Patents 381,968, 381,969, 381,970, 382,279, 382,280, 382,281, and 382,282 covered his single and polyphase motors, his distribution system, and polyphase transformers.
† The remainder of his polyphase system were numbered 390,413, 390,414, 390,415, 390,721, 390,820, 487,796, 511,559, 511,560, 511,915, 555,190, 524,426, 401,520, 405,858, 405,859, 406,968, 413,353, 416,191, 416,192, 416,193, 416,194, 416,195, 445,207, 459,772, 418,248, 424,036, 417,794, 433,700, 433,701, 433,702, 433,703, 455,067, 455,068, and 464,666.

Morgan he would soon be commuting in his private railway car—at first from Pittsburgh to New York but finally to Niagara Falls. In his reputation as a plunger, Westinghouse somewhat resembled Edison. And like Edison he was a fighter. The two men were to be well-matched in the battles ahead.

Westinghouse was a hard-driving businessman but he was the antithesis of a robber baron: he did not see the buying up of politicians and the fleecing of the public as essential to success in business. What he did see, what he had appreciated from the very first, was the potential of a power system that could send currents of high voltage surging across the great spaces of America. Like Tesla he had even dreamed of harnessing the hydroelectric potential of Niagara Falls.

He called on the inventor in his laboratory. The two men, who shared both the romance of the new energy and a taste for personal dandiness, felt a quick rapport. Tesla's workshops and laboratory were crammed with intriguing apparatus. Westinghouse moved from machine to machine, sometimes bent forward, hands on knees, peering, or sometimes with his head tilted, nodding with pleasure at the smooth hum of alternating-current motors. He needed few explanations.

There is a story, unfortunately without documentation, that he then turned to Tesla and offered him $1 million plus a royalty for all of his AC patents. If ever made, the offer must have been declined, for the records show that for his forty patents Tesla received about $60,000 from the Westinghouse firm, which included $5,000 in cash and 150 shares of stock. Significantly, however, according to Westinghouse historical records, he was to earn $2.50 per horsepower of electricity sold.* Within a few years these royalties would be worth such a stupendous amount of money that they would pose a curious problem.

For the present, however, since the monies received by Tesla had to be shared with Brown and other investors in his firm, he was far

---

* Memorandum of Agreement dated July 7, 1888, between Westinghouse Electric Company and Tesla Electric Company. A further agreement between Nikola Tesla and the Westinghouse Electric Company was signed July 27, 1889. Several earlier biographers incorrectly state that Tesla's royalty was to be only $1 per horsepower sold.

from having joined the super-rich. Nevertheless his transition from threadbare to fashionable in the social circles of Manhattan was both agreeable and slightly dizzying.

He agreed to work as a consultant for Westinghouse in adapting his single-phase system, at a salary of $2,000 per month. While the extra income was welcome, it meant moving to Pittsburgh just as exciting social invitations had begun to trickle in from members of the New York "400." He left reluctantly.

As might have been anticipated with a completely new system, difficulties lay ahead. The 133-cycle current then used by Westinghouse was wrong for Tesla's induction motor, which was built to 60 cycles. When he so informed the engineers, he succeeded in rubbing them the wrong way and only after months of futile and costly experiments doing it their way did they finally accept his word. Once they had done so, the motor worked exactly as it had been designed to do. Sixty cycles has ever since been the standard for alternating current.

Tesla soon achieved another milestone as important to him as the development of his inventions. On July 30, 1891, he became an American citizen. This, as he often told friends, he valued more than any of the scientific honors to come to him. Honorary degrees he tossed into drawers, but his certificate of naturalization was always kept in his office safe.

After several months he finished his duties in Pittsburgh and returned to New York, feeling physically and mentally exhausted. To a large extent he felt those months wasted since they had kept him from moving ahead with new research.

In September he left for Paris to attend the International Exposition and, from there, in the company of his uncle Petar Mandić, departed for Croatia. Petar had once been a monk in the monastery of Gomirje near Ogulin, and here the exhausted inventor went to recover his health.

He then visited his sisters and mother. Of the circumstances in which his widowed mother then lived or whether he ever contributed to her support once he began to earn money in America, unfortunately no records have been found. That she often dominated his thoughts, however, future events were to disclose.

Edison felt a flood of outrage when he first heard the news of Tesla's deal with Westinghouse for his alternating-current system. At last the lines were clearly drawn. Soon his propaganda machine at Menlo Park began grinding out a barrage of alarmist material about the alleged dangers of alternating current.[4] As Edison saw it, accidents caused by AC must, if they could not be found, be manufactured, and the public alerted to the hazards. Not only were fortunes at stake in the War of the Currents but also the personal pride of an egocentric genius.

By now the bad times had turned to boom. The country was expansion-minded. There were steelworks in Pittsburgh, a new Brooklyn Bridge, towers reaching toward the sky above Manhattan. Railroads, land, and gold were making fortunes for those who speculated in growth at the right time. Edison himself had become one of the leading industrialists in America, employing almost 3,000 workers at his various plants.

Michael Pupin, who later joined with Edison and Marconi to form a damaging trinity against his fellow Serb, was among those who immediately saw the superiority of Tesla's AC system. In fact he claimed that he came near to being fired from the electrical-engineering faculty at Columbia University for "eulogizing" this new technology.

Pupin, a farm boy who had grown up on the military frontier of Serbia, had arrived in New York at the age of fifteen with a nickel in his pocket (one cent more than Tesla), had shoveled coal for fifty cents a ton, and later won scholarships to Columbia University and Cambridge. Like Tesla he became one of America's greatest physicists and electrical engineers.

But it disturbed Pupin that the captains of the electrical industry were paying so little attention to highly trained electrical experts. All they seemed to worry about, he charged, was that their direct-current systems would not be supplanted by alternating current.

"A most un-American mental attitude!" said this new American. "It was clear to every impartial and intelligent expert that the two systems supplemented each other in a most admirable manner."

The patents held by Westinghouse were challenged by a number of litigants, primarily rival manufacturers claiming that their inventors had anticipated Tesla. Suits were filed in behalf of the inventors Walter Baily, Marcel Deprez, and Charles S. Bradley. In addition, in an attempt to evade the Tesla patents, General Electric filed an application for what was called the "monocyclic" system of their brilliant mathematician, Charles Steinmetz. Steinmetz himself, however, never questioned Tesla's preeminence in the AC field.

Such actions confused the public, and even some members of the engineering profession never clearly understood that the system almost universally adopted was Tesla's. This confusion is, to some extent, still true, despite the sweeping and eloquent ruling in Tesla's favor issued in September 1900 by Judge Townsend of the U.S. Circuit Court of Connecticut. If for no other reason than that, Judge Townsend's words are worth quoting here:

It remained to the genius of Tesla to capture the unruly, unrestrained and hitherto opposing elements in the field of nature and art and to harness them to draw the machines of man. It was he who first showed how to transform the toy of Arago into an engine of power; the "laboratory experiment" of Baily into a practically successful motor; the indicator into a driver; he first conceived the idea that the very impediments of reversal in direction, the contradictions of alternations might be transformed into power-producing rotations, a whirling field of force.

What others looked upon as only invincible barriers, impassable currents and contradictory forces he seized, and by harmonizing their directions utilized in practical motors in distant cities the power of Niagara.

A decree may be entered for an injunction and an accounting as to all the claims in suit.

At West Orange, New Jersey, families living in the neighborhood of Edison's huge laboratory began to notice that their pets were vanishing. Soon they found out why. Edison was paying schoolboys twenty-five cents a head for dogs and cats, which he then electrocuted in deliberately crude experiments with alternating current. At the same time he issued scare leaflets with the word "WARNING!" in red letters at the top. The gist of these messages: if the public were not alert, they might find themselves being terminally "Westinghoused."

Edison had been laying the groundwork for his vendetta for two years. He had written to E. H. Johnson: "Just as certain as death Westinghouse will kill a customer within six months after he puts in a system of any size. He has got a new thing and it will require a great deal of experimenting to get it working practically. It will never be free from danger. . . ."[5]

Now he was accusing Westinghouse of doing what he himself had done to the gas companies when he sent agents around the country propagandizing the virtues of direct current: "None of his plans worry me in the least; only thing that disturbs me is that W. is a great man for flooding the country with agents and travelers. He is ubiquitous and will form numerous companies before we know anything about it. . . ."[6]

Westinghouse, his eyes on the challenges ahead, paid only reluctant attention to Edison's hectoring but at last he agreed to carry on an educational campaign to combat it. He would make speeches, he said; he would write articles; he would do anything to get the truth before the people. He was, he told Tesla, determined to win for his company the right to harness Niagara Falls.

He also had his eye on Chicago and the Columbian Exposition to be held there in 1893. Planners were already beginning to speak of this event—commemorating the 400th anniversary of America's discovery—as the World of Tomorrow, the White City that would light up the land. He could not have asked for a better showcase.

Unfortunately, Lord Kelvin, the famous English scientist, had been named chairman of the International Niagara Commission established to choose the best means of harnessing the Falls, and Kelvin had declared himself squarely on the side of old-fashioned direct current.

When the commission offered a prize of $3,000 for the most practicable plan, about twenty were submitted. But the Big Three electrical companies, Westinghouse, Edison General Electric, and Thomson-Houston, elected not to participate. The commission had been set up by a New York group called the Cataract Construction Company, the president of which was Edward Dean Adams. As Westinghouse saw it, this firm was "trying to get one hundred thousand dollars' worth of information for three thousand dollars." When they were "ready to talk business," he said, he would submit his plans.

As usual in these years of rapid growth, George Westinghouse had money problems. It had cost a great deal more than he had expected to convert his plants over to the Tesla polyphase system. And now when he needed funds for expansion, the bankers were giving him mingy responses.

His only consolation was knowing that Edison was in trouble too. The rumors on Wall Street were that, unless Edison consolidated, his problems were acute. To take his mind off them, he blustered. Westinghouse, he said, should stick with his air brakes, for he knew nothing about the electricity business.

Edison's opening feint in the War of the Currents was to lobby legislators at Albany to pass a law limiting electrical currents to 800 volts. That way, he figured, AC would be stopped. But the legislators didn't buy it since Westinghouse countered with a threat to sue the Edison firm and others for conspiracy under the laws of the State of New York.

"The man has gone crazy," ranted Edison of his nemesis in Pittsburgh, "and is flying a kite that will land him in the mud sooner or later." [7]

In addition to waging a virulent campaign in press, pamphlet, and by word of mouth, Edison initiated Saturday demonstrations for newspaper reporters with strong stomachs. He called them in to witness the frightened dogs and cats that schoolboys had snatched off the streets being shoved onto a sheet of metal to which were attached wires from an AC generator with a current of one thousand volts. [8]

Batchelor sometimes helped with these demonstrations of the perils of alternating current. Once while trying to hold a wriggling puppy, he himself received a terrible shock. He described having "the awful memory of body and soul being wrenched asunder . . . the sensations of an immense rough file thrust through the quivering fibres of the body." Still the killing of animals continued.

Edison was in this fight literally to the death, although not his own. He, Samuel Insull, and a former laboratory assistant named Harold P. Brown worked out a scheme to finish Westinghouse once and for all, or so they thought—through the death of a third party.

Brown managed by subterfuge to buy a license to use three of the Tesla AC patents without Westinghouse knowing of their intended purpose. Brown then made a trip to Sing Sing Prison. Shortly afterward the prison authorities announced that the death house would carry out future executions not by hanging but by electrocution, and more specifically by alternating current, courtesy of the Westinghouse patents.

Prior to the next execution "Professor" Brown went on the road with Edison's traveling show. On stage he electrocuted a number of calves and large dogs with AC and referred to having "Westinghoused" them. In effect he was asking Americans, "Is this the invention you want your little wife to cook dinner with?"

Public concern had been fired to the desired pitch when New York State prison authorities announced the first scheduled electrocution of a condemned murderer. One William Kemmler would die on August 6, 1890—Westinghoused.

Kemmler was strapped into the electric chair and the switch thrown. But Edison's engineers, all their experiments having been with smaller creatures, had erred. The electric charge was too weak, and the condemned man was only half-killed. The dreadful procedure then had to be repeated. A reporter described it as "an awful spectacle, far worse than hanging."[9]

Westinghouse through the long, sordid campaign doggedly continued to try to set the public straight about AC, citing facts and figures to support its safety. Luckily he had prestigious help from Professor Anthony at Cornell, Professor Pupin at Columbia, and other respected scientists.

Edison's associates eventually began to sense that the tide might be turning and tried to convince the great inventor that, from the standpoint of his own industrial future, he was making a monumental mistake. But stubbornness was one of his weaknesses, and he refused to see it. It would be twenty years before he would admit that this had been his greatest blunder. After all, one of his favorite sayings was: "I don't care so much for a fortune . . . as I do for getting ahead of the other fellow."

But long before Edison was prepared to admit scientific error, it was borne in on him that his priorities must be revised. His financial difficulties had grown extreme, and a merger seemed almost inevitable.

The Thomson-Houston Company provided an object lesson when it was taken over by the House of Morgan and placed under the direction of a professional manager named Charles A. Coffin. An apt student of J. Pierpont Morgan, Coffin waged price wars against his competitors and then, once they were weak, wheedled them into lethal mergers. Along the way, Thomson and Houston lost control of their firm.

As Westinghouse later described an interview with Coffin to Clarence W. Barron: "He [Coffin] told me how he ran his stock down and deprived both Thomson and Houston of the benefits of an increased stock issue. He was enabled, by the decline in stock which he had forced, to make a new contract with both Thomson and Houston, by which they waived their rights to take new stock in proportion to their holdings under their agreement with the Company.

"I said to Coffin, 'You tell me how you treated Thomson and Houston; why should I trust you . . . ?'"[10]

Edison, however, was not granted the luxury of deciding whether he trusted Coffin. On February 17, 1892, *The Electrical Engineer* announced a consolidation of the Edison Electric Company and the Thomson-Houston Company, with none of the founders' names in the new title. Henceforth the new firm would be called General Electric Company, with Coffin as its president.

In the same article *The Electrical Engineer* has written:

It seems quite reasonable to expect, as many do, and as rumor has it, that absorption of the Westinghouse Company into the proposed new corporation will soon follow. The provision of $16,600,000 of stock—$6,000,000 of which is in preferred shares—remaining to the treasury after taking up the Edison and Thomson-Houston stocks, is thought by many to imply the use of a considerable portion of it in taking over the Westinghouse Company when convenient; but no definite information of such a plan has been made public.

In short, Morgan was close to realizing his ambition of controlling the future electrification of America, both AC and DC, through the elimination of "costly competition." He meant to use the same tactics that had worked so well in centralizing the control of the railroads, of oil, coal, and steel. Clearly, the best growth investments of the future would be in controlling the manufacture of all electrical appliances and machinery and providing the related services that eventually would become known as "public utilities." But to do this he would need the Tesla patents.

Coffin, in his reckless interview with Westinghouse, had revealed that he had been "cutting prices fearfully" in order to "knock out" other electrical firms. The important thing, he advised, man-to-man, was to get one's own system installed before the competition did so, whether it be for running electric trolleys or whatever; after that, any changes would be prohibitively expensive. "The users willingly pay our price as they cannot afford to change the system," he exulted.[11] He had been talking to precisely the wrong person, however, for Westinghouse was committed to proving that a superior system could indeed knock out an entrenched but inferior one.

Coffin had spoken earnestly of the advantages of "boodle." He had asked Westinghouse to raise the price of his street lights from $6 to $8, as his own firm had done, since this would enable him to pay $2 in boodle to the aldermen and other politicians without losing a cent of profit.[12] But when it became clear that Westinghouse was not to be a willing partner to his own demise, General Electric Company and the House of Morgan turned upon him where he was most vulnerable, in the money markets.

"From all the stock-market sub-cellars and rat-holes of State, Broad, and Wall streets crept those wriggling, slimy snakes of bastard rumors," wrote Thomas Lawson in *Frenzied Finance*. "'George Westinghouse has mismanaged his companies . . . George Westinghouse . . . is involved beyond extrication unless by consolidation with the General Electric. . . .' There came a crash in the Westinghouse stocks."

Lawson reports that he himself was called in as "an expert in stock market affairs" to assist Westinghouse, and that he drove a heroic bargain. First, there must be a consolidation of some sort. Westinghouse was indeed overextended in his drive to put the country on an alternating-current system.

The financial advisers arranged a merger with several smaller companies including U.S. Electric Company and the Consolidated Electric Light Company. The new firm would be known as the Westinghouse Electric and Manufacturing Company.

So far, so good, but there was one problem: Nikola Tesla's patent royalties under the generous arrangement with Westinghouse would sink any ship, according to the investment bankers. One source has stated he was told by Tesla that Westinghouse had paid him $1 million in advance royalties.[13] Only four years after the contract was signed, it was rumored that the accrued royalties could be in the neighborhood of $12 million. No one seemed to know exactly, least of all Tesla. As utilities expanded, royalties would be collected on powerhouse equipment and motors and on every application of the alternating-current system patents. Tesla stood to become a billionaire, one of the world's wealthiest men.

"Get rid of that royalty contract, Westinghouse," the investment banker advised. Otherwise the stability of the reorganization would be emperiled.

This Westinghouse was loath to do. He himself was an inventor and believed in royalties. Besides, he argued, royalties were paid for by the customers and included in costs of production. But the bankers left him with no choice.

Reluctantly he called on the inventor in what must have been one of the most embarrassing confrontations of his life. (In the official biography of George Westinghouse the episode goes unmentioned.) The contract between Tesla and Westinghouse had been made in good faith on the part of both men. Tesla, had he chosen, undoubtedly could have gone to court and had it upheld. But to what end if Westinghouse were to lose his firm?

As usual, George Westinghouse went directly to the point. Explaining the problem, he said, "Your decision determines the fate of the Westinghouse Company."[14]

Tesla's absorption in his new fields of research had been total. Money was something he spent freely when he had it, but he seldom knew how much was available. To him the value of money consisted in what one did with it rather than in any intrinsic worth.

"Suppose," he asked, "I should refuse to give up my contract; what would you do then?"

Westinghouse spread his hands. "In that event you would have to deal with the bankers, for I would no longer have any power in the situation."

"And if I give up the contract, you will save your company and retain control? You will proceed with your plans to give my polyphase system to the world?"

"I believe your polyphase system is the greatest discovery in the field of electricity," said Westinghouse. "It was my efforts to make it available to the world that brought on the present difficulty. But I intend to continue, no matter what happens, with my original plans to put the country on an alternating-current basis."

Being no businessman, Tesla could not refute Westinghouse's assessment of his financial situation; but he trusted the industrialist. "Mr. Westinghouse," he said, "you have been my friend, you believed in me when others had no faith; you were brave enough to go ahead . . . when others lacked courage; you supported me when even your own engineers lacked vision to see the big things ahead that you and I saw; you have stood by me as a friend. . . . You will save your company so that you can develop my inventions. Here is your contract and here is my contract—I will tear both of them to pieces, and you will no longer have any troubles from my royalties. Is that sufficient?"[15]

The Westinghouse Company's annual report of 1897 states that Tesla was paid $216,600 for outright purchase of his patents at this point to avoid the payment of royalties.

By destroying the contract, Tesla not only relinquished his claim to millions of dollars in already earned royalties but to all that would have accrued in the future. In the industrial milieu of that or any other time it was an act of unprecedented generosity if not foolhardiness. He was to live well for another decade but thereafter would be plagued by a chronic shortage of research and developmental capital. How many discoveries were thus to be lost to society can only be surmised.

Westinghouse returned to Pittsburgh, where the mergers and refinancing were arranged. His company went on to become a giant, and he kept his promise to Tesla. Years later in a formal testimonial to the industrialist, Tesla wrote: "George Westinghouse was, in my opinion, the only man on this globe who could take my alternating-current system under the circumstances then existing and win the battle against prejudice and money power. He was a pioneer of imposing stature, one of the world's true noblemen of whom America may well be proud and to whom humanity owes an immense debt of gratitude."[16]

Tesla had returned from his months in Pittsburgh depressed not only by his disagreements with the Westinghouse engineers but because of several lawsuits just beginning over his alternating-current inventions.

"Hundreds of electrical manufacturers pirated the Tesla patents," John J. O'Neill noted in a private communication, "and when Westinghouse had them completely upheld in the courts and smashed down on the trespassers, all of the hate of the losers was vented on Tesla."

Some attacks went beyond simple piracy. Claims were advanced in behalf of Professor Galileo Ferraris of the University of Turin as the first to have described a method of producing a revolving magnetic field. He apparently had given some thought to the problem in 1885 but had made no progress. Tesla, by comparison, had made his discovery of the rotating magnetic field in 1882 and within two months had evolved the complete system, which included all the apparatus he later patented. He had actually built his first induction motor. Ferraris, however, had concluded that the principle could never be used for making a practical motor.

He nevertheless had been publicized by *The Electrician* in London as the man most likely to invent one. When the editors learned of Tesla's invention, they erroneously assumed and reported that he had been inspired by the concept of Ferraris.

Because of the vicious rivalry between Edison and Westinghouse, the former faction seized on any opportunity to smear Tesla. The specious argument over Ferraris seemed as good an excuse as any.

Two prominent immigrants (although they would later be allied with the Edison camp) rose at once to Tesla's defense. Steinmetz, in a paper for the American Institute of Electrical Engineers, said: "Ferraris built only a little toy, and his magnetic circuits, so far as I know, were completed in air, not in iron, though that hardly makes any difference."

As for Professor Michael Pupin, he wrote to Tesla: "The Ferraris humbug has been indulged in by your competitors to a disgraceful extent. As I understand it there is a gigantic step from Ferraris' whirling pool to Tesla's whirling magnetic field. The two things seem to me radically different and ought to be pointed out and shown in their true lights. . . ."[17]

Tesla, deep in his research, was scarcely aware of the corrosive antagonisms raging around his inventions. He was immersed in a whole new world of electrical phenomena.

Westinghouse, meanwhile, when he was not testifying or speechifying, was aggressively extending the front lines of his industrial domain. Out in the little mining town of Telluride, Colorado, the first commercial use was made of Tesla motors and generators built by Westinghouse. They were installed in 1891 to furnish electricity for the mining camps.[18]

6. ORDER OF THE FLAMING SWORD

As long as the world left him alone in his Manhattan laboratory to pursue his love affair with electricity, Tesla was the happiest man alive. In the waning years of the 1880s and the early 1890s he had enjoyed such a brief period. But when he delivered four blockbusting lectures in America and Europe in 1891–92, he became, in a matter of months, the world's most celebrated scientist, and his private life was never the same again.

A weird, storklike figure on the lecture platform in his white tie and tails, he was nearly seven feet tall, for he wore thick cork soles during his dangerous demonstrations. As he warmed to his act, his high-pitched, almost falsetto voice would rise in excitement. The audience, riveted by the cadenced flow of words, the play of lights and magic, would stare as in a trance.

The language of science then being completely inadequate, Tesla described visual effects in the style of a poet in love with the sheer dance of flame and light. Indeed it seemed as if these were as significant to him as tapping the energy within. Yet no scientist could fault him on technical details.

Despite the fireworks, philosophy, and poetry, his every scientific claim was based on experiments he had personally repeated at least twenty times. Each item of equipment was new, designed by him and usually fabricated in his own shop. The same demonstration was seldom repeated from one appearance to another.

As to the inadequacy of the scientific terminology of his day—the luminous feathery discharge of electricity in a vacuum tube that he referred to in his lectures as a brush was in fact a beam of electrons and ionized gas molecules. He did not say, "Now I shall describe the cyclotron," for the word was nonexistent; but what he would describe and what he would demonstrate was thought by some who were knowledgeable to have been an early ancestor of the atom smasher.

Nor did he say, "Now I shall describe the point electron microscope. Now I shall describe cosmic rays. Now I shall describe the radio vacuum tube. Now I shall describe X ray." When he described a vacuum bulb which turned out to be the forerunner of the Audion, radio was called wireless, and wireless was scarcely in its infancy. When he described blurred photographic plates in his laboratory, and visible and invisible light, not even Roentgen knew what X rays were or might be used for. And when Tesla created a flame that he described as "burning without consuming material or even a chemical reaction," he probably was venturing into plasma physics.

"Phenomena upon which we used to look as wonders baffling explanation, we now see in a different light," he told the American Institute of Electrical Engineers. "The spark of an induction coil, the glow of an incandescent lamp, the manifestations of the mechanical forces of currents and magnets are no longer beyond our grasp; instead of the incomprehensible, as before, their observation suggests now in our minds a simple mechanism, and although as to its precise nature all is still conjecture, yet we know that the truth cannot be much longer hidden, and instinctively we feel that the understanding is dawning upon us. We still admire these beautiful phenomena, these strange forces, but we are helpless no longer. . . ."[1]

He spoke of the mysterious fascination of electricity and magnetism, "with their seemingly dual character, unique among the forces in nature, with their phenomena of attractions, repulsions, and rotations, [their] strange manifestations of mysterious agents," that stimulate and excite the mind.

But how to explain them?

"An infinitesimal world, with the molecules and their atoms spinning and moving in orbits, in much the same manner as celestial bodies, carrying with them and probably spinning with them ether, or in other words, carrying with them static charges," he said, "seems to my mind the most probable view, and one which, in a plausible manner, accounts for most of the phenomena observed. The spinning of the molecules and their ether sets up the ether tensions or electrostatic strains; the equalization of ether tensions sets up other motions or electric currents, and the orbital movements produce the effects of electro and permanent magnetism."

It had been only three years since, speaking before this same professional group, he had introduced the power system that was to revolutionize industry and bring light to even the most remote homes. Now he described his research into the very nature of electricity by way of light and luminous effects, holding his audience in thrall.

The stage from which he spoke was illuminated with stunning displays of gas-filled tube lights, some of which had been made phosphorescent to enhance their brilliance and for some of which he used uranium glass. They were the forerunners of today's fluorescent lights. Tesla never patented or commercialized them, and they did not appear on the market until fifty years later. For his lecture, characteristically, he had twisted the tubes into names—not only those of great scientists but of his favorite Serbian poets.

Turning to a table, the spellbinder carefully selected a delicate prop. "Here is a simple glass tube from which the air has been partially exhausted," he said. "I take hold of it; I bring my body in contact with a wire conveying alternating currents of high potential, and the tube in my hand is brilliantly lighted. In whatever position I may put it, wherever I move it in space, as far as I can reach, its soft, pleasing light persists with undiminished brightness."[2]

As the tube he held began to glow—demonstrating among other things a political message about the safety of alternating current— "Professor" Brown, the Edison agent, arose unnoticed and hurried from the hall. His boss would chew nails when he heard about this razzle-dazzle. But George Westinghouse, who had come from Pittsburgh just to hear the lecture, leaned forward, shook his head, and smiled.

Tesla next revealed his wireless or electrodeless discharge lamps inductively coupled to a high-frequency power supply, which he had invented after discovering that gases at reduced pressure exhibited extremely high conductivity. These, as he showed, could be moved anywhere in the room yet would eerily continue to burn. He would never get around to making them practicable for commercial use, but they are still being investigated more than eighty years later, as shown by patents recently issued.

Roland J. Morin, the chief engineer of Sylvania GTE International, New York, later wrote: "I am sure that [Tesla's] demonstration of these light sources at the Chicago World's Fair [1893] stimulated D. McFarlan Moore to develop and announce commercial realization of the fluorescent lamp. . . ."

Gracious in paying tribute to scientists who had paved the way, Tesla expressed his debt to Sir William Crookes, who in the 1870s had built a vacuum tube with a pair of electrodes inside. Alluding to "that same vague world" (later identified as a stream of electrons), he spoke of the effects obtained with alternating currents of high voltage and frequency: "We observe how the energy of an alternating current traversing the wire manifests itself—not so much in the wires as in the surrounding space—in the most surprising manner, taking the forms of heat, light, mechanical energy and, most surprising of all, even chemical affinity."

His long fingers deftly chose another prop.

"Here is an exhausted bulb suspending from a single wire. . . . I grasp it, and a platinum button mounted in it is brought to vivid incandescence.

"Here, attached to a leading wire, is another bulb which, as I touch its metallic socket, is filled with magnificent colors of phosphorescent light.

"Here again," he said, "insulated as I stand on this platform, I bring my body in contact with one of the terminals of the secondary of this induction coil . . . and you see streams of light break forth from its distant end, which is set in violent vibration. . . .

"Once more, I attach these two plates of wire gauze to the terminals of the coil. [T]he passage of the discharge ... assumes the form of luminous streams."

It was impossible with an induction coil, he said, to pursue any novel investigation without coming upon some interesting or useful fact. He began to describe effects he had achieved in the laboratory—"large pin-wheels, which in the dark present a beautiful appearance owing to the abundance of the streams," and of how he had sought to produce "a queer flame which would be rigid."

To his listeners it sometimes seemed as if visual excitement were as important to him as useful results; but then, in the next breath he would present them one "useful fact" after another.

For example, he showed them a motor that ran on only one wire, the return circuit occurring wirelessly through space. And, renewing his spell over men who prided themselves on common sense and resistance to flimflammery, he spoke of the possibility of running motors without any wires at all. He spoke of energy in space, free for the taking.

"It is quite possible," he said, "that such 'no-wire' motors, as they might be called, could be operated by conduction through the rarefied air at considerable distances. Alternating currents, especially of high fre-quencies, pass with astonishing freedom through even slightly rarefied gases. The upper strata of the air are rarefied. To reach a number of miles out into space requires the overcoming of difficulties of a merely mechanical nature. There is no doubt that with the enormous potentials obtainable by the use of high frequencies and oil insulation, luminous discharges might be passed through many miles of rarefied air, and that, by thus directing the energy of many hundreds of thousands of horse-power, motors, or lamps might be operated at considerable distance from stationary sources. But such schemes are mentioned merely as possibili-ties. We shall have no need to transmit power in this way. We shall have no need to *transmit* power at all. Ere many generations pass, our machin-ery will be driven by a power obtainable at any point of the universe. This idea is not novel. ... We find it in the delightful myth of Antaeus, who derives power from the earth; we find it among the subtle specula-

tions of one of your splendid mathematicians. . . . Throughout space
there is energy. Is this energy static or kinetic? If static, our hopes are in
vain; if kinetic—and this we know it is, for certain—then it is a mere
question of time when men will succeed in attaching their machinery to
the very wheelwork of nature. . . ."[3]

The star exhibit of Tesla's show, however (to be elaborated on in his
later lectures in England and France), was a single six-inch almost empty
vacuum tube that he called the carbon-button lamp. With this research
tool he explored whole new areas of scientific discovery.[4]

It was a small glass globe with a tiny piece of solid material mounted
on the end of a wire serving as a single-wire connection with the high-
frequency current source. The central "button" of material electrostatically propelled the surrounding gas molecules toward the glass globe.
They then were repelled back toward the button, striking it and heating
it to incandescence as the process occurred millions of times each second.

Depending on the strength of the source, extremely high temperatures could be produced that vaporized or melted most substances
instantly. Tesla experimented with buttons composed of diamonds,
rubies, and zirconia. He finally found that carborundum did not vapor-
ize as rapidly as other hard materials or make deposits inside the globe—
hence the name, carbon-button lamp.

The heat energy of the incandescent button was transferred to the
molecules of the slight amount of gas in the tube, causing them to
become a source of light twenty times brighter for the amount of energy
consumed than Edison's incandescent lamp.

With hundreds of thousands of volts of high-frequency currents
surging through his body, he held in his hand this magnificent creation, a
working model of the incandescent sun. With it he demonstrated what
he believed to be cosmic rays. The sun, he reasoned, is an incandescent
body carrying a high electrical charge and emitting showers of tiny particles, each of which is energized by its great velocity. But, not being
enclosed in a glass, the sun permits its rays to strike out into space.

Tesla was convinced that all space was filled with these particles, constantly bombarding Earth or other matter, just as in his carbon-button lamp the hardest material was shattered into atomic dust.

One of the manifestations of such bombardment, he said, was the aurora borealis. Although no record exists of his methods, he announced that he had detected such cosmic rays, measured their energy, and found them moving with a velocity of hundreds of millions of volts.[5]

The more sober physicists and engineers in his audience, hearing such outrageous claims, kept their counsel. But where was the evidence?

Today it is known that thermonuclear reaction on the sun causes the radiation of X rays, ultraviolet, visible, and infrared rays as well as radio waves and solar particles at the rate of 64 million watts (or volt-amperes) per square meter of the sun's surface.

Cosmic rays, according to modern knowledge, come in many shapes and forms and are the result of the formation and decay of particles as well as the high-energy collision of particles. They come not only from the sun but from the stars and novae or exploding stars.

Solar electrons and protons reaching the vicinity of the Earth and trapped by Earth's magnetic field form the Van Allen radiation belts. Solar radiation, visible and invisible, determines the surface temperatures of the planets. Auroral displays are caused by solar-emitted particles that collide with the atoms in our upper atmosphere.

Five years after Tesla's lecture Henri Becquerel, the French physicist, was to discover the mysterious rays emitted by uranium. Marie and Pierre Curie confirmed his work with their study of radium, whose uranium atoms were exploding spontaneously. Tesla had believed, wrongly, that cosmic rays were the simple cause of the radioactivity of radium, thorium, and uranium. But he was entirely correct in predicting that bombardment with "cosmic rays," i.e., high-energy subatomic particles, could make other substances radioactive, as was finally demonstrated by Irene Curie and her husband Frédéric Joliot in 1934.

Although the scientific world of Tesla's time did not accept his theory of cosmic rays, two scientists who later achieved fame in this field would acknowledge a debt to his inspiration. Thirty years were to pass before Dr. Robert A. Millikan rediscovered cosmic rays. He believed them to be, like light, vibratory—that is, that they were photons rather than charged particles. This led to one of the scientific dog fights of the 1940s between Nobel laureate Millikan and Nobel laureate Arthur H. Compton, who believed—and indeed was adjudged to have proved—that cosmic rays consisted of high-velocity particles of matter, just as Tesla had described them.

Both Millikan and Compton expressed their debt to the intuitiveness of their Victorian predecessor. But science was to march on inexorably, proving cosmic rays more varied and complex than any of them had guessed.

The strange little carbon-button lamp with which Tesla dazzled his audience at Columbia College on May 20, 1891, also embodied the concept of the point electron microscope. It produced electrified particles shooting out in straight lines from a tiny active spot on the button, kept at high potential. On the spherical surface of the globe these particles reproduced in phosphorescent images the pattern of the microscopically tiny area from which they were issuing.[6]

The only limit to the degree of magnification that could be obtained was the size of the glass sphere. The greater the radius, the greater the magnification. Since electrons are smaller than light waves, objects too tiny to be seen by light waves may nevertheless be enlarged by the patterns produced by emitted electrons.

Vladimir R. Zworykin is credited with having developed the electron microscope in 1939. Yet Tesla's description of the effect achieved with his carbon-button lamp when he used extremely high vacuum stands with hardly a change in wording for a description of the million-magnification point electron microscope.[7]

Another effect produced by the carbon-button lamp derived from the phenomenon of resonance. In describing the principle of resonance, Tesla often used analogies of a wine glass and a swing. A wine glass that is broken by a violin's note is shattered because the vibrations of the air that are produced by the violin happen to be of the same frequency as the vibrations of the glass.

A person in a swing may weigh two hundred pounds and a weak boy pushing it may weigh but fifty and may push but a pound. Yet if he times his pushes to coincide with the turn of the swing from him, and keeps adding a pound each time, he will eventually have to stop to avoid hurling the occupant of the swing out into space.

"The principle cannot fail," Tesla would say. "It is necessary only to keep adding a little force at the right time."

And that is why Tesla's carbon-button lamp may be described as an ancestor of the atom-smasher. Using the hard carborundum button in a nearly air-exhausted globe, connecting it to a source of high, rapidly alternating current, he caused the remaining molecules of air to become charged, thus to be repelled at increasingly high velocities from the button to the glass globe, and thence back to the button, shattering the carbon beads in the button into atomic dust which joined the oscillating air molecules to cause even further disintegration.

"If the frequency could be brought high enough," he said, "the loss due to the imperfect elasticity of the glass would be entirely negligible...."[8]

In 1939 Ernest Orlando Lawrence of the University of California, Berkeley, won the Nobel Prize for his invention of the cyclotron. According to one account: "In 1929, Ernest Orlando Lawrence ... read a paper by a German physicist who had managed, by giving two electrostatic impulses instead of one, to impart to charged potassium atoms in a vacuum tube twice the energy they would normally get from a given voltage. Lawrence wondered: if the impulse could be doubled, could it not be tripled or multiplied any number of times? The problem was to give the particles a series of impulses a little stronger each time, until, like a child being pushed on a swing, the momentum was greatly increased."[9]

He made a particle-pushing machine of glass and sealing wax. The disk-shaped vacuum chamber was only four inches wide. Inside were two electrodes, each shaped like half a round cake box and called

D plates. Outside the vacuum chamber was a powerful electromagnet. Electrified particles or protons were whirled in a magnetic field in the circular chamber until they attained very high speed and were then fired out of the chamber in a narrow stream of high-speed atomic bullets. Lawrence's first model was called a cyclotron because it whirled the protons in circles. Soon he built a larger one that fired protons up to energies of 1.2 million electron volts.

Whether Tesla was actually smashing the carbon's atomic nucleus, as his first biographer thought, has little bearing on the revolutionary nature of his achievement. The inventor himself described the molecules of the residual gas as violently impinging on the carbon button and causing it to rise to an incandescent state, or a near-plastic phase of the solid.

Lawrence may have had no knowledge of Tesla's molecular-bombardment lamp. Undoubtedly, however, he did know of the attempts to build an atom-smasher that were made by Gregory Breit and his associates at the Carnegie Institution in Washington, D.C., in 1929, for this group used a 5-million-volt Tesla coil to supply the necessary power. Without such apparatus, the machines necessary to crack the atom could never have functioned.

The descriptions of Tesla's carbon-button or molecular-bombardment lamp are to be found in the permanent records of five learned societies.* Unfortunately in the early 1890s no society was sufficiently learned to imagine a use for this technological ancestor of the Atomic Age.

Frédéric and Irene Joliot-Curie, Henri Becquerel, Robert A. Millikan, Arthur H. Compton, and Lawrence all won Nobel prizes. Victor F. Hess won the Nobel in 1936 for discovering cosmic radiation. Surely it would be an act of simple justice were the scientific community at least to acknowledge Tesla's pioneer discoveries in each of their fields.

Although many—perhaps most—of his scientific contemporaries failed to understand his lectures fully, Tesla fired the imaginations of a

* AIEE, Columbia College, May 20, 1891; Institution of Electrical Engineers and Royal Society of Great Britain, London, February 1892; Society of Electrical Engineers of France and the French Society of Physics, Paris, February 1892. For books containing his lectures, see the bibliography.

perceptive few. And like some today who discover him for the first time, a kind of temporary madness seized them. "Not only did he teach by accomplishment," recalled Maj. Edwin H. Armstrong, who later won fame for his contributions to radio, "but he taught by the inspiration of a marvelous imagination that refused to accept the permanence of what appeared to others to be insuperable difficulties: an imagination the goals of which, in a number of instances, are still in the realms of speculation."[10]

The English scientist J. A. Fleming wrote Tesla: "I congratulate you most heartily on your *grand* success. . . . After that no one can doubt your qualifications as a magician of the first order. Say the Order of the Flaming Sword."[11]

To trace Tesla's productivity in a sequential fashion in this period is almost impossible. He seemed everywhere at once, working in a dozen fields that overlapped and were interrelated—but always with electricity, that mysterious substance, at the heart of his investigations. To him it was a fluid with transcendental powers that condescended to obey certain physical laws, rather that a stream of discrete particles, or wave packets obeying certain particle laws, as in modern theory.

Nevertheless, in the next few years he was to disclose the whole direction of modern electronics, although the electron itself would not be discovered until 1897 by the British physicist Joseph J. Thomson.

Faraday had shown in 1831 that it was possible to convert mechanical energy into electric current. Then, in the year of Tesla's birth, England's Lord Kelvin had made a further discovery that would inspire the Serbo-American when he began seeking a new source of high-frequency currents, higher than could be mechanically produced.

It had been believed that when a condenser was discharged, the electricity flowed out of one plate into the other like water. Kelvin showed that the process was complex, that electricity rushed from one plate into the other and back again until all the stored energy was used up, surging at a tremendously high frequency of hundreds of millions of times a second.

On the day in Budapest when the concept of the rotating magnetic field was revealed to Tesla, he had seen in a flash the universe composed of a symphony of alternating currents with the harmonies played on a vast range of octaves. The 60-cycles-per-second AC was but a single note in a lower octave. In one of the higher octaves at a frequency of billions of cycles per second was visible light. To explore this whole range of electrical vibration between his low-frequency alternating current and light waves, he sensed, would bring him closer to an understanding of the cosmic symphony.

The work of James Clerk Maxwell in 1873 had indicated the existence of a vast range of electromagnetic vibrations above and below visible light—vibrations of much shorter and much longer wavelengths. This theory had been tested by Prof. Heinrich Hertz of Germany who, in a search for waves longer than light or heat, first produced man-made electromagnetic radiation at Bonn in 1888. Hertz's experiments with the spark discharge of an induction coil had proved the existence of a magnetic field when he sent a powerful electric charge across a spark gap, causing a smaller spark to jump across a second gap some distance away. At the same time in England, Sir Oliver Lodge was seeking to measure tiny electrical waves in wire circuits.

Hertz's equipment had been feeble and the spark coil both impractical and dangerous. Tesla now came up with something both different and very much superior: a series of high-frequency alternators producing frequencies up to 33,000 cycles per second (33,000 Hz.).* This type of machinery was in fact the forerunner of the great high-frequency alternators developed by others for continuous-wave radio communication in the distant future, but for the inventor's immediate needs, the device was still inadequate. He therefore went on to build what is known as the Tesla coil, an air-core transformer with primary and secondary coils tuned to resonate—a step-up transformer which converts relatively low-voltage high current to high-voltage low current at high frequencies.

This device for producing high voltages, which is today used in one form or another in every radio and television set, was in a very short time

---

* Today this would be in the medium-to-low range.

to become part of the research equipment of every university science laboratory. It allowed the operator to convert the weak, highly damped oscillations of the original Hertz circuit and to sustain currents of almost any magnitude. In this research Tesla thus anticipated by several years the first experiments of Marconi.

The need for insulating this high-voltage equipment led to his immersing it in oil to exclude all air, a method that soon found valuable commercial application, since it became the universal way of insulating all high-tension apparatus. To reduce resistance in his coils, Tesla used stranded conductors with separately insulated strands. Since he seldom took time to patent his research tools or methods, this too went into the common pool of knowledge. It was later commercialized by others, becoming known as "Litz wire," a term derived from *Litzendraht* ("stranded wire") cable.

He then developed a new kind of reciprocating dynamo adapted to his special needs in high-frequency currents—an ingenious single-cylinder engine without valves, that could be operated by compressed air or steam. The speed it attained was so remarkably constant that he proposed adapting it to his 60-cycle polyphase system, using synchronous motors, properly geared down, as a means of providing the correct time wherever in the world alternating current was available. This was the inspiration for the modern electric clock.[12] Tesla, in his rush of discovery, took no time to patent a timekeeper either.

And not least, from the dangerous experiments in which he learned to work with hundreds of thousands of volts of high-frequency electricity came another discovery of great importance to the world. In 1890 he announced the therapeutic deep-heating value of high-frequency currents on the human body. The process became known as diathermy. From it would flow an enormous field of medical technology, with many early imitators both in America and Europe.[13]

7. RADIO

Long hours of mental exertion in his New York laboratory over many months caused Tesla to experience a strange partial amnesia at the start of the 1890s.

Immediately after finishing his consulting work for the Westinghouse corporation, he had become obsessed with what was first spoken of as the wireless telephone—or simply wireless—and later by its modern name, radio.

After building the powerful coils in his laboratory, he had ascertained that broadcasting intelligence was simply one aspect of a vast global and interplanetary potential. Radio posed a different set of problems from transmitting electricity without wires, yet he believed them close enough to be tackled in a single stunning orchestration.

"I had produced a striking phenomenon with my grounded transmitter," he later recalled, "and was endeavoring to ascertain its true significance in relation to the currents propagated through the earth. It seemed a hopeless undertaking, and for more than a year I worked unremittingly but in vain. This profound study so entirely absorbed me that I became forgetful of everything else, even of my undermined health. At last, as I was on the point of breaking down, nature applied the preservative, inducing lethal sleep."[1]

From having gone almost without rest for months, he said that he had then slept "as if drugged." On regaining his senses, he was shocked to discover that he could visualize no scenes from his past except those of earliest infancy.

Having developed a marked indifference to medical doctors, he put his mind to the problem of curing himself.

Night after night he concentrated on the memories of early childhood, gradually bringing more and more of his life into focus. In this

unfolding process the image of his mother was always the principal figure. He began to feel a consuming desire to go to her.

"This feeling grew so strong," he recalled, "that I resolved to drop all
work and satisfy my longing. But I found it too hard to break away from
the laboratory and several months elapsed, during which I succeeded in
reviving all the impressions of my past life. . . ."

It was early spring of 1892. He had not yet accepted a flock of invitations to lecture in England and France and indeed was still in a state of
emotional conflict about doing so.

Then, he recalls, a vision materialized "out of the mist of oblivion,"
and he saw himself at the Hotel de la Paix in Paris, just coming to from
one of his peculiar sleeping spells. In this "recollection," he saw himself
being handed a dispatch bearing the sad news that his mother was dying.

A curious fact about this period of partial amnesia, Tesla later wrote,
was that he was alive to everything touching on his research, which went
forward apace. "I could recall the smallest details and the least significant
observations in my experiments, and even recite pages of texts and complex mathematical formulae."

It appears there had been reason for his concerns about his mother's
health: letters had been arriving from the family home at Gospic indicating that her health was indeed failing. He had also been receiving from all
parts of the world invitations, honors, "and other flattering induce-
ments" to visit and lecture. At last he accepted those from London and
Paris, planning thereafter to go directly home.

His lecture to the Institution of Electrical Engineers in London was
hailed as a major scientific event, and when it was over, the British did not
want to let him go.

"Sir James Dewar insisted on my appearing before the Royal
Society," he recalled. "I was a man of firm resolve but succumbed easily
to the forceful arguments of the great Scotchman. He pushed me into a
chair and poured out half a glass of wonderful brown fluid which
sparkled in all sorts of iridescent colors and tasted like nectar."

To his surprise Dewar said, "Now you are sitting in Faraday's chair and you are enjoying whiskey he used to drink."[2] On being assured that no one else in the world more deserved these honors, he was won over. The French could wait one more day.

His lecture before the Royal Society of Great Britain, attended by the elite of the scientific world, brought yet more accolades for the young inventor. Lord Rayleigh, the distinguished physicist who was then chairman of the Royal Society, urged the inventor, because of his great talent for mining fundamental discoveries, to consider revising his modus operandi.

He recommended that Tesla in the future specialize in some single area of research. This was a highly novel idea for a scientist who demanded all the answers at once.

Sir William Crookes, whose work Tesla greatly admired, sent a letter to his hotel after the lecture, describing how he had been inspired to subject his own body to strange electrical effects.

"My dear Tesla," he wrote. "You are a true prophet. I have finished my new coil, and it does not do so well as the little one you made for me. I fear it is too large.... The phosphorescence through my body when I hold one terminal is decidedly inferior to that given with the little one...."[3]

The observant Crookes had noted the inventor's exhaustion and went on to warn him that he appeared to be on the verge of a physical and nervous breakdown. "I hope you will get away to the mountains of your native land as soon as you can," he wrote. "You are suffering from overwork, and if you do not take care of yourself, you will break down. Don't answer this letter or see anyone but take the first train."

Sir William was right; but his advice was just then impossible for Tesla to accept.

The inventor hurried to Paris where he lectured on "Experiments with Alternating Currents of High Potential and High Frequency" and again demonstrated his sensitive electronic tubes. This time his audiences were the Société Internationale des Electriciens and the Société Française de Physique.

That same month of February 1892, Sir William Crookes affirmed

Tesla's intuition. He published a prediction that electromagnetic waves in space could be used for wireless.

No sooner had Tesla finished his last lecture than, pleading exhaustion, he fled to his room at the Hotel de la Paix. It seemed almost an anticlimax when a messenger handed him a telegram saying that his mother was dying.

Rushing to the station, he squeezed aboard a train just leaving for Croatia. Later transferring to a carriage, he reached home in time to spend a few hours with his mother. Then, near collapse, he was taken to a building close to his home to rest.

"As I lay helpless there," he wrote in his autobiographical memoir, "I thought that if my mother died while I was away from her bedside she would surely give me a sign. . . . I [had been] in London in company with my late friend, Sir William Crookes, when spiritualism was discussed, and I was under the full sway of these thoughts. . . . I reflected that the conditions for a look into the beyond were most favorable, for my mother was a woman of genius and particularly excelling in the powers of intuition."[4]

During that entire night his mind was strained with expectancy, but nothing happened until early in the morning. In a light dream or "swoon," he says, he saw "a cloud carrying angelic figures of marvelous beauty, one of whom gazed upon me lovingly and gradually assumed the features of my mother. The appearance slowly floated across the room and vanished, and I was awakened by an indescribably sweet song of many voices. In that instant a certitude, which no words can express, came upon me that my mother had just died. And that was true. . . ."

It was important to him later to account for the external causes of these apparently trancendental impressions, since he still held to his thesis that human beings were mere "meat machines." The following "explanation" appears in his memoir:

"When I recovered I sought for a long time the external cause of this strange manifestation and, to my great relief, I succeeded after many months of fruitless effort. I had seen the painting of a celebrated artist, representing allegorically one of the seasons in the form of a cloud with a group of angels which seemed to actually float in the air, and this had struck me forcefully. It was exactly the same that appeared in my dream, with the exception of my mother's likeness. The music came from the choir in the church nearby at the early mass of Easter morning, explaining everything satisfactorily in conformity with scientific facts.

"This occurred long ago, and I have never had the faintest reason since to change my views on psychical and spiritual phenomena, for which there is absolutely no foundation. The belief in these is the natural outgrowth of intellectual development. Religious dogmas are no longer accepted in their orthodox meaning, but every individual clings to faith in a supreme power of some kind. We all must have an ideal to govern our conduct and insure contentment, but it is immaterial whether it be one of creed, art, science or anything else, so long as it fulfills the function of a dematerializing force. It is essential to the peaceful existence of humanity as a whole that one common conception should prevail.

"While I have failed to obtain any evidence in support of the contentions of psychologists and spiritualists, I have proved to my complete satisfaction the automatism of life, not only through continuous observations of individual actions, but even more conclusively through certain generalizations." [5]

He said that whenever friends or relatives of his had been hurt by others in a particular way, he himself felt what he could only characterize as a "cosmic" pain. This resulted from the fact that human bodies are of similar construction and exposed to the same external influences, which results in likeness of response. "A very sensitive and observant being, with his highly developed mechanism all intact, and acting with precision in obedience to the changing conditions of the environment," he wrote, "is endowed with a transcending mechanical sense, enabling him to evade perils too subtle to be directly perceived. When he comes in contact with others whose controlling organs are radically faulty, that sense asserts itself and he feels the 'cosmic' pain. . . ." [6]

It is obvious from the inventor's writings that he himself was never completely satisfied with his theories on this subject.

This was not to be the only instance of precognition and extrasensory perception in Tesla's life. But he always tried to explain them away mechanistically, tracing intuition to external events. Thus when his sister Angelina fell fatally ill, he sent a telegram from New York saying, "I had a vision that Angelina was arising and disappearing. I sensed all is not well." Tesla's nephew, Sava Kosanović would later recall how the inventor told him of such premonitions but discounted them. He was a sensitive receiver, he said, registering any disturbance—but there was no mystery to it.

"He declared," said Kosanović, "that each man is like an automaton which reacts to external impressions." But what the external impressions were that gave him actual *pre*cognition, as hereafter described, he never discussed.

He told Kosanović of an incident that occurred in Manhattan in the 1890s after he had given a big party. Some of the guests were preparing to take a train for Philadelphia. Tesla, seized with "a powerful urge," was impelled to detain them, causing them to miss the train. It crashed. Many passengers were injured.[7]

He associated a personal anomaly with the anxious rush to his mother's deathbed. A patch of white hair developed on the right side of his head, which was otherwise jet black and thick. After a few months, however, it returned to its natural state.

Following his mother's death he was ill for a number of weeks. When finally able to get about, he visited relatives in Belgrade, where he received the welcome due a world-famous native son, and then went on to Zagreb and Budapest.

As a child Tesla had been fascinated by the relationship between lightning and rain. On this trip, while roaming in his native mountains, he had an experience that profoundly affected him as a scientist.

"I sought shelter from an approaching storm," he later recalled. "The sky became overhung with dark clouds but somehow the rain was delayed until, all of a sudden, there was a lightning flash and a few moments after, a deluge. This observation set me thinking. It was manifest that the two phenomena were closely related, as cause and effect, and a little reflection led me to the conclusion that the electrical energy enclosed in the precipitation of the water was inconsiderable, the function of lightning being much like that of a sensitive trigger.

"Here was a stupendous possibility of achievement. If we could produce electric storms of the required ability, this whole planet and the conditions of existence on it could be transformed. The sun raises the water of the oceans and winds drive it to distant regions where it remains in a state of the most delicate balance. If it were in our power to upset it when and wherever desired, this mighty life-sustaining medium could be at will controlled. We could irrigate arid deserts, create lakes and rivers and provide motive power in unlimited amounts."

Controlling lightning, he concluded, would be the most convenient way of harnessing the power of the sun.

"The consummation depended on our ability to develop electric forces of an order of those in nature," he decided. "It seemed a hopeless undertaking, but I made up my mind to try. [I]mmediately upon my return to the United States, in the summer of 1892, work was begun which was to me all the more attractive, because a means of the same kind was necessary for the successful transmission of energy without wires."[8]

On August 31, 1892, *The Electrical Engineer* reported the return to New York of Mr. Nikola Tesla, the distinguished electrician, on the steamship *Augusta Victoria* from Hamburg. After commenting on the death of Tesla's mother and his subsequent illness, the journal added: "His magnificent reception at the hands of European electricians has become, like his investigations and researches, part of electrical history; and the honors conferred on him were such as to make Americans very proud of one who has chosen this country as a home."

He moved scientific history forward again in the spring of 1893 when, addressing the Franklin Institute in Philadelphia and the National

Electric Light Association at St. Louis, he described in detail the principles of radio broadcasting.

At St. Louis he made the first public demonstration ever of radio communication, although Marconi is generally credited with having achieved this feat in 1895.

Tesla's twenty-eight-year-old assistant at the St. Louis lecture was H. P. Broughton, whose son, William G. Broughton, is licensee of the Schenectady Museum memorial amateur radio station W21R. At the station's dedication speech in 1976 William Broughton touched upon highlights of Tesla's historic demonstration at St. Louis—after a week's preparation—as personally told to him by his father.

"Eighty-three years ago, in St. Louis, the National Electric Light Association sponsored a public lecture on high-voltage high-frequency phenomena," said the younger Broughton. "On the auditorium stage a demonstration was set up by using two groups of equipment.

"In the transmitter group on one side of the stage was a 5-kva high-voltage pole-type oil-filled distribution transformer connected to a condenser bank of Leyden jars, a spark gap, a coil, and a wire running up to the ceiling.

"In the receiver group at the other side of the stage was an identical wire hanging from the ceiling, a duplicate condenser bank of Leyden jars and coil—but instead of the spark gap, there was a Geissler tube that would light up like a modern fluorescent lamp bulb when voltage was applied. There were no interconnecting wires between transmitter and receiver.

"The transformer in the transmitter group," Broughton continued, "was energized from a special electric power line through an exposed two-blade knife switch. When this switch was closed, the transformer grunted and groaned, the Leyden jars showed corona sizzling around their foil edges, the spark gap crackled with a noisy spark discharge, and an invisible electromagnetic field radiated energy into space from the transmitter antenna wire.

"Simultaneously, in the receiver group, the Geissler tube lighted up from radio-frequency excitation picked up by the receiver antenna wire.

"Thus wireless was born. A wireless message had been transmitted by the 5-kilowatt spark transmitter, and instantly received by the Geissler-tube receiver thirty feet away. . . .

"The world-famous genius who invented, conducted, and explained this lecture demonstration," he concluded, "was Nikola Tesla."

Although the St. Louis demonstration was no "message sent 'round the world" as Tesla would doubtless of course have preferred it to be, he had nevertheless demonstrated all the fundamental principles of modern radio: 1. an antenna or aerial wire; 2. a ground connection; 3. an aerial-ground circuit containing inductance and capacity; 4. adjustable inductance and capacity (for tuning); 5. sending and receiving sets tuned to resonance with each other; and 6. electronic tube detectors.[9]

In his earliest transmissions he used vibrating contacts to make continuous waves in a receiving system audible. A few years later the crystal detector was introduced to receive the signals of spark-gap transmitters. This became the accepted practice of commercial radio until the invention by Maj. Edwin H. Armstrong of the regenerative or feedback circuit, which brought radio into the era of amplified sound. Later, Armstrong introduced the superheterodyne beat-note circuit, which underlies all modern radio and radar reception. Armstrong, a graduate student of Prof. Michael Pupin's at Columbia University, had been inspired by Tesla's lectures. Later, however, perhaps influenced by Pupin, he would champion Marconi in the prolonged and bitter war between the latter and Tesla over radio patents.

The scientist who, next to Tesla, most deserved credit for pioneering radio was Sir Oliver Lodge, for in 1894 he demonstrated the possibility of transmitting telegraph signals wirelessly by Hertzian waves a distance of 150 yards.

Two years later young Marchese Guglielmo Marconi arrived in London with a wireless set identical to Lodge's. Naturally he aroused little comment among the leading contenders in the race. He did, however, have a ground connection and antenna or aerial wire with which he had made crude experiments in Bologna. As it happened, this equipment was exactly what Tesla had described in his widely published lectures of 1893, which had been translated into many languages.[10] Later, as we shall see, Marconi was to deny that he had ever read of Tesla's system, and the U.S. Patent Examiner was to brand his denial patently absurd.

Significantly, until the early 1960s only eleven patent cases would reach the United States Supreme Court and of those few, two involved Tesla patents. The fundamental nature of his work was characteristic. The high court heard cases involving his polyphase alternating-current patents and his radio patents, and both actions were decided in his favor. Ironically, neither of these was brought by the inventor himself.

January sleet scraped at the windows of Tesla's laboratory. Kolman Czito, his assistant, shivered as he helped to adjust a machine, but the inventor worked away in total concentration. For all that Tesla was aware of the temperature, it might as well have been blossom time.

The telephone rang, and he sighed as he went to answer it. The operator was putting through a long-distance call from Pittsburgh.

George Westinghouse's voice boomed across the miles, almost stuttering in his excitement. His firm had gotten the contract for installing all the power and lighting equipment for the Chicago World's Fair of 1893, otherwise known as the Columbian Exposition—the first electrical fair in history. It would use Tesla's alternating-current system, his maligned and ridiculed AC, all the way.

This was good news and bad: good because it offered a great international event as a showcase; bad because it meant leaving work that meant more to him than anything else in life. His radio research was now at its most exciting, critical point.

The industrialist's words were tumbling over each other. It was going to be the grandest spectacle of modern times, he said; a chance not only to show what AC could do but to exhibit all the new electrical products being invented. Who would not give an arm and a leg for such an opportunity?

General Electric would be showing Edison's inventions. Everybody who was anybody in international science would be there. The architecture was to be magnificent.

"When does the Fair open?" Tesla asked, fearing the worst.

"May first. Hardly time for everything we must do."

"All right, Mr. Westinghouse," said the inventor.

Turning away from his beloved coils, he went to work on the big show. Ideas were already racing through his mind for ways to amaze the scientific community and bewitch the public. He could not possibly have said no.

The United States both wanted and needed a spectacle. Shortly after President Grover Cleveland was elected to a second term of office, the nation was engulfed by bank failures, joblessness, and bankruptcies. The Panic of 1893 haunted the humble and the mighty alike. Something to take people's minds off the imminent prospect of standing in breadlines seemed politically desirable.

The Columbian Exposition was devised as a celebration (one year late) of the four hundredth anniversary of the discovery of America. President Cleveland invited the royalty of Spain and Portugal and other foreign dignitaries. He even agreed to turn the gold master key that would release the electricity and flood the City of Tomorrow with light, starting up fountains and machinery, raising flags and banners, and signaling the grand opening of the extravaganza. To agree to turn the master key took courage. Electricity had been installed in the White House in 1891, but thus far no president had ever been allowed to touch the switches. The task had been prudently left to hirelings, for, after all, the public had been warned by no less an authority than Edison of the dangers involved.

Chicago was a gray city when the great day finally came, the breadlines now being actual and long. But the site of the Fair was breathtaking to the multitude that arrived, and reporters began to write of it as the White City. *The New York Times* (May 1, 1893) reported, "Grover Cleveland, calm and dignified, in a few eloquent words delivered in a clear, ringing voice, which was heard by the great multitude gathered before him, declared the World's Columbian Exposition open . . . and touched the ivory-and-gold key. . . ."

A Tower of Light flared into brilliance with a thousand electric bulbs radiating the promise of a brighter future. Venetian canals had been built to mirror the modern illumination of "Old World" architecture. Everywhere the pulse of the future throbbed: alternating current.

As the lights went on, the massed human beings below uttered a great sigh. Then, in the seats reserved for them, the Cabinet officers, the Duke and Duchess of Veragua, and other foreign dignitaries began to cheer. The crowd lustily joined in while tightly corseted women fainted and fell like soldiers in battle.

Westinghouse, who had underbid General Electric on the illumination contract, had enjoyed a decisive triumph. In the Electricity Building could be seen all the latest products and inventions of American ingenuity. At night especially the Fair seemed an enchanted place. Colored searchlights played on the fountains, making them so beautiful that people actually wept tears of joy. Adventurous citizens careered around the fairgrounds on an elevated train driven by electricity. The foolhardy crowded to get seats on Mr. G. W. Ferris's enormous wheel, which was 250 feet in diameter and like nothing ever seen before. They packed in sixty to a car to soar out precariously above both the White City and the gray city that lay beyond.

Between May and October, 25 million Americans visited Chicago to see the latest wonders of science, industry, art, and architecture. This was then a third of the total population.

Visitors crowded into the display rooms presided over by the famous Nikola Tesla. Clad in white tie and tails, he stood among a magician's feast of high-frequency equipment, demonstrating one electrical miracle after another. A darkened alcove held tables that glowed with his phosphorescent tubes and lamps. One length of tubing radiated the words "Welcome, Electricians," which Tesla had had laboriously blown letter by letter from the molten glass. His other lights honored such great scientists as Helmholtz, Faraday, Maxwell, Henry, and Franklin. And he had not forgotten—right up there with the famous scientists—the name of the most eminent living poet of Yugoslavia: Zmaj Jovan, whose pseudonym was ZMAJ.

Day after day he captivated the curious with demonstrations illustrating how alternating current worked.[11] On a velvet-covered table small metallic objects—copper balls, metal eggs—were made to spin at great speeds, reversing themselves smoothly at fixed intervals.

He demonstrated the first synchronized electric clock attached to an oscillator and showed his first disruptive discharge coil. The audiences understood little of the science involved, yet were enthralled. And when he seemed to turn himself into a human firestorm by using the apparatus with which he had so often thrilled his laboratory visitors, they cried out in fear and wonder.

A bevy of Tesla's young women friends arrived under firm escort from New York City. They flirted with him, rode on the Ferris wheel, and visited the Woman's Building to hear Mrs. Potter Palmer (Chicago's retort to Mrs. Astor) declare that the model kitchen, which boasted an electric stove, electric fans, and even an automatic dishwasher, heralded the liberation of the female.

It is possible, however, that they felt more liberated by the sight of Princess Eulalia who, representing her nephew King Alfonso of Spain, brazenly smoked cigarettes in public.

They saw the first zipper and Edison's Kinetoscope (early motion-picture photography) which brought "scenes to the eyes as well as sounds to the ear"; and they listened to thin bursts of music piped by telephone from a concert in Manhattan. They stood with crowds ogling the bellydancing of an energetic young woman billed as Little Egypt and—because the Fair offered something for every taste—admired a plump Venus de Milo molded in chocolate.

A journalist, one of a throng who visited the Tesla exhibition, sent this report to his newspaper:

"Mr. Tesla has been seen receiving through his hands currents at a potential of more than 200,000 volts, vibrating a million times per second, and manifesting themselves in dazzling streams of light. . . . After such a striking test, which, by the way, no one has displayed a hurried inclination to repeat, Mr. Tesla's body and clothing have continued for some time to emit fine glimmers or halos of splintered light. In fact, an actual flame is produced by the agitation of electrostatically charged molecules, and the curious spectacle can be seen of puissant, white, ethereal flames, that do not consume anything, bursting from the ends of an induction coil as though it were the bush on holy ground."

The inventor, it was reported, expected one day to envelop himself in a complete sheet of lambent fire that would leave him quite uninjured. Such currents, he claimed, would keep a naked man warm at the North Pole, and their use in therapeutics was but one of their practical possibilities.

"My first announcement [of medical diathermy] spread like fire and experiments were undertaken by a host of experts here and in other countries," he later wrote. "When a famous French physician, Dr. d'Arsonval, declared that he had made the same discovery, a heated controversy relative to priority was started. The French, eager to honor their countryman, made him a member of the Academy, ignoring entirely my earlier publication. Resolved to take steps for vindicating my claim, I went to Paris, when I met Dr. d'Arsonval. His personal charm disarmed me completely and I abandoned my intention, content to rest on the record. It shows that my disclosure antedated his and also that he used my apparatus in his demonstrations. . . ."[12]

Although Tesla is credited with having first recorded the fact (in 1891) that heat production resulting from the bombardment of tissue with high-frequency alternating currents could have medical uses for the treatment of arthritis and many other afflictions, the name "D'Arsonval current" persisted in medical terminology. In any event, the use of radiation spread rapidly, and a field of medical technology—at first called diathermy and now called hyperthermia—developed that today includes the application of X rays, microwaves, and radio waves to destroy cancer cells. They are also used for healing bones and tissue.

Throughout his life Tesla was also a firm believer in the therapeutic value of what he called "cold fire," both for refreshing the mind and cleansing the skin. In fact, the brush discharge or corona from a low-power therapeutic device does seem to enhance muscular action, may improve circulation, and also generates ozone, which can be mildly stimulating when breathed in low concentration. Physicist Maurice Stahl said, "There is also a psychosomatic effect. I would consider the overall effect more than mechanical."

The inventor also had hopes that electrical anaesthesia might become possible. And he proposed burying high-voltage wires in classrooms to stimulate dull students. To key up actors before they went on stage, he arranged to install a high-tension dressing room in a New York theater.

At the Columbian Exposition Tesla also described heating bars of iron and melting lead and tin in the electromagnetic field of specially designed high-frequency coils. This was to have important commercial consequences many years later.

Although he had left his laboratory for Chicago reluctantly, the Fair proved an exhilarating experience for him. It was equally so for George Westinghouse. The latter displayed in the Machinery Hall various commercial motors of the AC system and twelve generators of the two-phase type that had been built especially for distributing light and power. To show the complete adaptability of his system, Westinghouse demonstrated how a rotary converter could change polyphase AC into DC to run a railway motor.

Perhaps Tesla's biggest day came on August 25 when he delivered a lecture to the Electrical Congress and demonstrated his mechanical and electrical oscillators. Thomas Commerford Martin, the well-known editor and electrical engineer, wrote that scientists would now be able to carry on investigations in alternating current with great precision. But also, he added, one of the uses of such equipment would be in the field of "harmonic and synchronous telegraphy" and that "vast possibilities are again opened up."[13]

Hermann Helmholtz, the celebrated German physicist, attended the Electrical Congress as an official delegate of the German Empire and was elected its president. Tesla's fellow countryman Michael Pupin, too, was a participant. "The subjects discussed at that congress," Pupin later wrote, "and the men who discussed them, showed that the electrical science was not in its infancy, and that electrical things were not done by the rule of thumb." Thus he too repudiated Edison's contention that alternating current was too little understood for safe use.

Tesla returned to New York elated by his triumphs. In the flush of fame he was more determined than ever to avoid the many public claims on his time. He would have preferred to avoid all commercial claims as well, but the need to finance radio and other research soon made this impossible.

8. HIGH SOCIETY

Wall Street was dominated by personal adventures, including such legendary figures as Morgan, John D. Rockefeller, the Vanderbilts, Edward H. Harriman, Jay Gould, Thomas Fortune Ryan, and other more ephemeral but equally colorful specimens. Some might bloom for a day, only to be trampled and forgotten. Most thrived on trading of such dubious legality that anyone who tried to emulate them today would probably be obliged to live in a foreign capital beyond threat of extradition. Dealing in coal, railroads, steel, tobacco, and the new field of electrical utilities, they plunged, cornered, and sold short.

According to the irreverent Twain, the gospel as preached by the robber barons during this galloping phase of the industrial revolution was, "Get money. Get it quickly. Get it in abundance. Get it dishonestly, if you can, honestly, if you must."

Each day when the closing gong sounded at the Stock Exchange on Wall Street, many members moved on to the Waldorf-Astoria Hotel, which was then located where the Empire State Building now stands. For a broker to be admitted as a member of the "Waldorf Crowd" was a patent of success. The splendid lounges and dining rooms served as showcases in which to observe the preenings of the winners as well as the dismay of the losers. Fear was often a palpable presence.

Tesla instinctively gravitated to the glass-enclosed Palm Room to see and be seen by the money men so important to his career. He had begun dining there regularly some years before he was able to take up residence at the fashionable hotel. Compared to the enormous wealth amassed by the plungers and builders of the period, he was not affluent, but he was handsome, polished, charming, and lived as if his prospects for wealth were excellent, as indeed they were. And after all, as Ward McAllister observed of the Gilded Age, "A man with a million dollars can be as happy nowadays as though he were rich."

Tesla himself was now a member of McAllister's exclusive roster of wealth and social position, the New York "400." He was meeting those fabled "great silent men with cold eyes and hard smiles" on their own playing field. His knowledge was being courted, and he enjoyed the game. Should he allow himself to become, like Edison, "Morganized"? Should he be "Astored," "Insulled," "Melloned," "Ryaned," or "Fricked"? He had no illusions as to the risk involved. No matter who capitalized his inventions, there would certainly be meddling interference and probably ultimate control. That was how the system worked, and it was the price an inventor must pay.

A few knowledgeable men had already begun to call him the greatest inventor in history, greater even than Edison. If further proof of his success in the New World were needed, a backlash was developing against him—not just in the Edison camp but, more quietly, among other scientists who received less attention from the press and who were never invited to the exciting celebrity affairs in his laboratory.

All his life Tesla was to cultivate an adoring host of journalists, editors, publishers, and literati. Although his lectures made him world-famous and were preserved in the records of learned societies, he never once submitted an article to an academic journal. Indeed, when he first arrived in America there was none; institutional ties with the big three of industry, government, and universities had not yet become the accepted avenue to recognition for a scientist. But now that was changing.

He was a loner by preference when the time for lone operators was swiftly passing. Edison himself, as one of the last of the "independents," was a transitional figure who built the first of the large industrial research laboratories, setting the style for modern science.

Tesla's lifelong distaste for corporate involvement was twofold: most other engineers drove him mad with impatience, and he resented any form of control. If he had to deal with a corporate person, he preferred it to be the president or chairman of the board.

The movers and shakers he observed at the Waldorf after the Stock Exchange had closed for the day were limited conversationalists. Their interests were largely in rates and tariffs, their fears riveted on financial panics and labor riots. Partisan politics scarcely interested them, only the buying of blocs of votes as necessary to protect the rates and tariffs. Bernard Baruch once told of a crude German trader named Jacob Field, known as Jake, who was being wined and dined by some grateful friends. When two lovely women on either side of him were stumped to know what to talk to him about, one of them finally asked whether he liked Balzac. Jake tugged on his mustaches. "I never deal in dem outside stocks," he said.[1]

Journalists and bluestockings were much more to Tesla's natural taste. As for the gentlemen of the press, they were so enthralled by his charismatic presence that they could scarcely remember after meeting him whether he had bushy black hair or wavy brown, or what was the color of his eyes or the length of his thumbs—the latter being, curiously, a matter of intense interest.

Male writers of the period often affected a florid prose style of which Julian Hawthorne, the novelist and only son of Nathaniel Hawthorne, was a leading exponent. Smitten by his first meeting with Tesla, he described something like a vision seen in an opium den:

"I saw a tall, slender young man with long arms and fingers, whose rather languid movements veiled extraordinary muscular power. His face was oval, broad at the temples, and strong at the lips and chin; with long eyes whose lids were seldom fully lifted, as if he were in a waking dream, seeing visions which were not revealed to the generality. He had a slow smile, as if awakening to actualities, and finding a humorous quality in them. Withal he manifested a courtesy and amiability which were almost feminine, and beneath all were the simplicity and integrity of a child. . . . He has abundant wavy brown hair, blue eyes and a fair skin. . . . To be with Tesla is to enter a domain of freedom even freer than solitude, because the horizon enlarges so. . . ."[2]

On the other hand, one of the inventor's secretaries, as if reciting "Peter Piper," wrote that he had bushy black hair brushed back briskly.

Yet everyone seemed agreed as to the power of Tesla's personality. Franklin Chester in the *Citizen* (August 22, 1897) wrote that no one could look upon him without feeling his force. Chester described him as well over six feet tall (actually he was a towering six feet six inches), with large hands and abnormally long thumbs, "a sign of great intelligence." As to the inventor's controversial hair, Chester said it was straight, a deep and shining black, brushed sharply from over his ears to make a ridge with serrated edges. His cheekbones were high and Slavic, his eyes blue and deeply set, burning like balls of fire.

"Those weird flashes of light he makes with his instruments," Chester continued, "seemed almost to shoot from them. His head is wedge shaped. His chin is almost a point. . . . When he talks you listen. You do not know what he is saying, but it enthralls you. . . . He speaks the perfect English of a highly educated foreigner, without accent and with precision. . . . He speaks eight languages equally well. . . ."

Hearst's flamboyant editor Arthur Brisbane found the inventor's eyes "rather light," as a result of straining his mind so much. (Tesla claimed this was true.) Brisbane shared the prevailing view that long thumbs meant a powerful intellect, referring his readers to the very small thumbs of apes. Tesla's mouth, however, he thought too small and his chin, although not weak, not strong enough. He guessed his height at more than six feet and his weight at less than 140 pounds, and reported that he tended to stoop. Tesla's voice he described as being somewhat shrill, probably from psychic tension.

"He has that supply of self-love and self-confidence that usually goes with success."[3]

John J. O'Neill, the Pulitzer Prize–winning science editor of the New York *Herald Tribune*, who was to become Tesla's first biographer and devoted friend of many years, described his eyes as gray-blue, which he felt to be a matter of genetic inheritance rather than mental strain. To him Tesla was a god whose ethereal brilliance "created the modern era."[4]

From the romantic point of view, O'Neill noted, he was too tall and slender to pose as the physical Adonis, but his other qualifications more than compensated.

"He was handsome of face, had a magnetic personality, but was quiet, almost shy; he was soft spoken, well educated and wore clothes well."

As to Tesla's own view of such matters, he fancied himself as being the very best-dressed man on Fifth Avenue. Moreover, as he once told his secretary, he intended to remain so. His usual streetwear included a black Prince Albert coat and a derby hat, and these he wore in the laboratory too unless some important experiment demanded formal evening wear. His handkerchiefs were of white silk rather than linen, his neckties sober, and his collars stiff. He threw out all accessories, including gloves, after a very few wearings. Jewelry he never wore and felt strongly about as the result of his phobias.

Robert Underwood Johnson, shortly after meeting Tesla, arranged that he be given an honorary degree from his own alma mater, Yale University. And later when Columbia also conferred an honorary degree upon him, Johnson was called upon to describe the special virtues of the inventor's character. Tesla, he said, had a personality of "distinguished sweetness, sincerity, modesty, refinement, generosity, and force. . . ."

Women were smitten as often as his male admirers.

Miss Dorothy F. Skerritt, his secretary of many years, attested that even in old age his presence and manner were impressive. "From under protruding eyebrows," she wrote, "his deep-set, steel gray, soft, yet piercing eyes, seemed to read your innermost thoughts . . . his face glowed with almost ethereal radiance. . . . His genial smile and nobility of bearing always denoted the gentlemanly characteristics that were so ingrained in his soul."[5]

His friend Hawthorne was struck not only by Tesla's physical attractiveness but by his richness of culture. Seldom did one meet a scientist or engineer, he noted, who was also a poet, a philosopher, an appreciator of fine music, a linguist, and a connoisseur of food and drink.

"[W]hen it was question of the vintage of a wine, or the condition and cooking of an ortolan, he knew that, too." And when he spoke, Hawthorne claimed, one could read the future in his face, seeing "mankind . . . arise a Titan, and grasp the secrets of the skies. I saw a coming time when the race would no longer be forced to labor for the means of livelihood, when the terms rich and poor would no longer mean difference of material conditions, but of spiritual capacity and ambition; a time . . . even, when knowledge should be derived from sources now hardly imagined. . . ."[6]

Tesla displayed occasional streaks of cruelty that seemed motivated by likes and dislikes of an almost compulsive sort. Fat people disgusted him, and he made little effort to conceal his feelings. One of his secretaries was in his opinion too fat. Once she awkwardly knocked something off a table and he fired her. She pleaded with him on plump knees to change his mind but he refused to do so. He had a favorite joke about two of his ancient aunts that centered on the fact that both were sublimely ugly.

He could be equally imperious about his subordinates' clothes. A secretary might spend half a month's earnings on a new dress, and he would criticize it, ordering her to go home and change it before delivering a message to one of his important banker friends.

His employees seemed never to question his assumed role as an arbiter of taste and in fact were singularly loyal to him. Other qualities compensated. His assistants Kolman Czito and George Scherff, his secretaries Muriel Arbus and Miss Skerritt, stayed with him through thick times and thin. When he grew old and rambling, journalists would protect him from his own utterances. The science writers Kenneth M. Swezey and O'Neill, mere teenagers when they met him, came to worship him almost as a god. Hugo Gernsback, the famous science editor and a father of science fiction, would publish everything he could get of Tesla's, considering him at least as important as Edison.

This strangely captivating figure was to be courted not only by writers, industrialists, and financiers, but by musicians, actors, kings, poets, university trustees, mystics, and crackpots. Honors would be showered upon him; foreign governments would seek his services. People were to call him a wizard, a visionary, a prophet, a prodigal genius, and the greatest scientist of all time. But that was not all.

Some called him a faker and a charlatan, just as at times they defamed Edison when he too "went public" with his inventions and bragged precipitously to the press. Fellow scientists in the universities would never forgive Tesla this sin. Edison's fame outlived such charges, for he took the wise precaution of acquiring a fortune and power as well as a vast popular following. But Tesla's dollars would slip away like sand, and he would have to stand alone, aloof and indifferent to public opinion.

One harsh critic, Waldemar Kaempffert, science editor of *The New York Times*, was to brand him "an intellectual boa constrictor" in whose coils such innocents as J. P. Morgan and Colonel Astor had been as helpless prey. Kaempffert would describe him as a "medieval practitioner of black arts ... as vague as an oriental mystic," and accuse him (mixing the historical metaphor) of being a hopelessly retrograde Victorian, unable to accept the new atomic science of the twentieth century. His fellow journalists, sniffed Kaempffert, "though they could not understand what [Tesla] was talking about, were enthralled with his proposals to communicate with Mars and to transmit power without wires over vast distances."[7] And he strongly intimated that among the duped journalists was his opposite number on the *Herald Tribune*. O'Neill gave Tesla far too much credit, Kaempffert said, as a result of adolescent hero worship. O'Neill had met Tesla while working as a page in the New York Public Library and allegedly wrote poems to him. Kaempffert's attitude was perhaps explained by the following incident described by O'Neill:

In 1898 Tesla made a celebrated demonstration in Madison Square Garden of a remotely controlled robot boat and torpedoes. Kaempffert, then a student at City College, brashly engaged the famous scientist in conversation.

"I see how you could load an even larger boat with a cargo of dynamite," he volunteered, "cause it to ride submerged, and explode the dynamite whenever you wished by pressing the key just as easily as you can cause the light on the bow to shine, and blow up from a distance by wireless even the largest of battleships."

Tesla snapped back, "You do not see there a wireless torpedo. You see there the first of a race of robots, mechanical men which will do the laborious work of the human race."[8]

Envious scientists and critical journalists were not to be the only sources of Tesla's travail. Occultists seemed attracted to him, and odd men and women preoccupied with even stranger matters flocked to his banner, proclaiming him their very own Venusian. He had been born on Venus, they insisted, and arrived on Earth either by spaceship or on the wings of a large white dove.[9]

These unwelcome followers believed him to be a man of prophecy and great psychic power who "fell to Earth" to uplift ordinary mortals through the development of automation. Partly to discourage all who would attribute abnormal powers to him, Tesla went to great lengths to deny even the sensory gifts he actually possessed. In the same spirit, he went farther than that, expounding his mechanistic philosophy, proclaiming that human beings were without wills of their own, their every act the result of external events and circumstances.[10] Despite all his disclaimers, however, the strange champions continued to follow him, sometimes linking his name with unfortunate publicity schemes. Who but a charlatan, it was asked, would attract such people?

One autumn evening Tesla's hansom cab deposited him at the fashionable home of the Robert Underwood Johnsons' at 327 Lexington Avenue. Arc lights sparkled in the frosty air as cabriolets, broughams, and other smart carriages delivered a careful assortment of guests. From the opened door drifted the strains of a Mozart piano concerto. The Johnsons were not wealthy, but they evenhandedly accumulated millionaires, supermillionaires, poor artists, and intellectuals. Neither Robert nor Katharine understood much about science but they both adored Tesla for his varied charms.

They were an attractive couple, he scholarly in appearance with a gift for languages, poetry, and repartee, Katharine petite and pretty, yet too intelligent and restless to be satisfied with her wife-and-mother role.

In addition to cultivating artists, they were genuinely interested in the arts. Johnson was the associate editor of *Century* magazine and later became its editor. Their home became a natural haven for the cultured Tesla, who missed the civilized rituals of Old World cities. Both he and Michael Pupin, although they came from the poorest backgrounds in Yugoslavia, had been appalled when they first confronted the vulgar clamor of America. At the Johnson home Tesla met prominent Continental artists, writers, and political figures as well as the cream of American society.

He was introduced to the Johnsons in 1893 by Thomas Commerford Martin, and liked them immediately. Soon the trio became fast friends. With Robert and Katharine, Tesla learned to relax his formal manners, to use first names, and even to relish the gossip of the times. Tesla's relentless search for millionaires to finance his inventions became the subject of the trio's favorite in-joke.

When they were not together, they exchanged notes—sometimes two or three times a day—by messenger. Over the years their correspondence amounted to thousands of letters between Robert and Nikola, but almost equally between Katharine and "Mr. Tesla," as she unfailingly addressed him even when her notes made no effort to conceal the intensity of her feelings for him. It was not long until Tesla loosened up enough to give them nicknames, calling Johnson "Luka Filipov" after a legendary Serbian hero he admired, and Mrs. Johnson "Madame Filipov." Johnson, in return, took up the study of Serbian.

The invitations from the Johnsons to Tesla convey an idea of the frenetic social life the inventor was leading at this time. "Do drop in if you can on your way to the Leggett's from the Van Allen's. . . ." "Come meet the Kiplings," "Come see Paderewski," "Come to meet Baron Kaneko. . . ." In his acceptances, Tesla sometimes signed his notes to the "Filipovs" with such frivolous names such as Nicholas I. or the initials

"G.I." (for Great Inventor). With few other friends did he feel able to be so playful.

Thanks to the Johnsons, Tesla was now also being given access to those special preserves of privilege where the Idle Rich played at the game of life with such single-minded ostentation and vulgarity. Robert described for him the banquets given at Delmonico's by the fabulously wealthy. They were called Silver, Gold, and Diamond dinners, depending upon which kind of jewelry was to be tucked into the napkins to surprise the women guests. Sometimes, for a taste thrill, cigarettes made of hundred dollar bills were passed around and smoked.

And if he did not attend it, the inventor most certainly read in the society pages of the bizarre soiree called the Poverty Social. The event in question was given in the brownstone mansion of a western hides-and-tallow king. Guests were commanded to show up in dirty rags. They sat on a filthy floor, swilling beer from tin cans and eating scraps of food served to them by liveried footmen, on wooden plates. Sensitivity was not one of the hallmarks of the Gilded Age.

But questions of taste aside, wealth had its undeniable attractions. "The only way I shall ever have a cent," said Tesla, "is when I have enough money to throw it out of the window in handfuls."[11]

At this time he was living at the Gerlach, which declared itself on its letterhead to be a "strictly fireproof family hotel." He chafed in these unglamorous surroundings and dreamed of the Waldorf on Fifth Avenue with its heavily gold-embossed stationery.

At the Johnson home, in addition to being introduced to Rudyard Kipling, whom he and Robert considered one of the great poets of the age, he met the writers John Muir and Helen Hunt Jackson, the composers Ignace Paderewski and Anton Dvořák, the prima donna Nellie Melba, and a parade of socialites and politicians, including Senator George Hearst.[12] He also met an unknown but strikingly handsome southerner just graduated from the U.S. Naval Academy, Richmond Pearson Hobson.

Tesla was already thirty-seven and a cosmopolitan, not easily impressed by new acquaintances. But he felt curiously attracted to the young officer whose boyish features were in such absurd contrast to the dark, swashing mustache he affected. Hobson was to come as close as any Serbian hero to Tesla's ideal—the virile, romantic man of action who combined native intelligence with a cultured background.

Among the animadversions against Tesla were whispers that he was a homosexual. In another time or another country it might have made little difference to his career; but in Victorian America, in the sober company of engineers, such rumors were to become a virulent part of the arsenal of his enemies. Since he could never be bothered to repudiate gossip of any kind, at any time, the only explanation he ever cared to advance for his celibacy was the exclusive demands imposed by his work. This, however, was unacceptable to the society of the time, and the pressures upon him to marry were unrelenting.

On the face of it, Tesla's phobias made him an unlikely candidate for intimate relationships. He did, however, at one period maintain an apartment at the luxurious Hotel Marguery on the west side of Park Avenue between 47th and 48th streets at the same time that his residence was at another hotel; and he once told Kenneth Swezey that he used it for meeting "special" friends and acquaintances. The statement, however, is open to many interpretations.

The Johnsons introduced him to a parade of women who were comely, talented, or rich, and sometimes all three. A fair number were said to be sexually attracted to him. He never responded in kind, but such attentions obviously gratified his ego.

On the autumn evening when he arrived at the Johnson home to hear the strains of Mozart drifting from the door, he recognized the pianist as Marguerite Merington, one of his perennial favorites as a dinner partner. The admiration and affection he felt for her appeared to be as much as he was ever to feel for any woman.

He was taken by Johnson to meet a tall, serious girl wearing an expensive French gown, modishly cinched in at the waist, with lace and a flower at the neckline. As she turned, her tawny eyes startled him. He was sure he had not met her, yet he had seen those eyes. An actress, perhaps?

"Miss Anne Morgan," said Johnson. "Mr. Tesla." Then he left them.

She nodded and returned her attention to the music. Tesla was amused. Of course. Her eyes had the same bold intelligence as her father's. He could almost visualize her lighting up a black cigar. Johnson had said the girl was in love with him. If so, she seemed determined not to betray it. Her poise, cultivated at so-called dames' schools, impressed him. So rich and yet so lovely.

What a pity, though, that the girl wore pearl earrings; they almost set his teeth on edge. He would have enjoyed talking with her, but the pearls made it impossible. Perhaps Robert would be kind enough to drop her a hint for the future. According to Elisabeth Marbury, Anne had been so overprotected as to be almost pathetically childlike. But if Tesla was any judge, the self-possessed creature before him would very soon be shedding her cocoon. Her metamorphosis would be interesting to watch.

The Johnsons, as he realized, were bound to tease him if he did not promptly display an interest in marrying the daughter of J. Pierpont Morgan. For an ambitious inventor in need of capital, he recognized the pitfalls in the situation. He could not decently encourage the young woman in her infatuation, but he must be extremely diplomatic to avoid hurting her feelings.

When the music ended, others claimed his attention. At parties these days he was always quickly surrounded. People longed to hang upon the words of the gifted spellbinder. The wealthy tended not to be scientifically critical, and Tesla relieved their boredom. He in turn enjoyed letting his fancies fly.

On the evening in question he made an excuse and sought out Marguerite whose candor he appreciated. Complimenting her on her performance, he asked somewhat tactlessly, "Tell me, Miss. Why do you not wear diamonds and jewelry like the others?"

"It is not a matter of choice with me," she said. "But if I had enough money to load myself with diamonds, I could think of better ways of spending it."

"What would you do with money if you had it?" he asked with interest.

"I would prefer to purchase a home in the country, except that I would not enjoy commuting from the suburbs."[13]

Tesla beamed. Fancy a charming and talented woman who rejected jewels. He himself never wore even a tiepin or a watch chain.

"Ah, Miss Merington, when I start getting my *millions*," he said, "I will solve that problem. I will buy a square block here in New York and build a villa for you in the center and plant trees all around it. Then you will have your country home and will not have to leave the city."[14]

She laughed, briefly wondering, perhaps, if this were some kind of proposition. But it is unlikely she could have concluded that Tesla's words were anything but banter.

According to one of the inventor's close friends, Marguerite later claimed to be the only woman who ever touched Tesla. The friend discounted it. No record of intimacy linking her or any other woman to the inventor has ever been discovered.

The same confidante said that Anne Morgan "threw herself" at Tesla. Again there is nothing to support the belief that they were more than friends. They were to enjoy parallel careers, Anne becoming a most vital and important woman in her own right. Although her name would be linked with a succession of famous men, she would never marry.

Periodically to repay his social debts Tesla gave elaborate banquets at the Waldorf for members of the "400" and lesser mortals. Invitations were jealously sought for these splendid affairs. He personally selected the choicest foods and liquors, supervised their preparation, hovered over the sauces, and anguished over the vintage wines. No cost was spared, and no plebeians were invited.

After such affairs the guests were titillated by visits to his laboratory for private "showings," and many a prophetic announcement appeared in the next day's papers about his exciting inventions. He could not have chosen a more telling way to torment those of his scientific contemporaries who were excluded from such performances.

But still his relative indifference to women continued to be a subject of international gossip. One night as he sat in the Café de la Paix in Paris

with a French scientist, a theater party passed that included the divine
Sarah Bernhardt. The actress coyly dropped her handkerchief. He
sprang to his feet and returned it to her without so much as raising his
eyes and at once, to the dismay of the Frenchman, resumed his discussion of electricity.

Even the *Electrical Review* of London (August 14, 1896) devoted a
lengthy editorial to chiding him: "Of course Mr. Tesla may be quite
invulnerable to Cupid's shafts, but somehow or other we doubt it. We
are great admirers of him and his work, and we give him credit for good
hard sense. . . . We have faith enough in women to believe that his fate will
come, and that some one will be found who is not only a match for his
intensity in all respects, but who will tax his inventive genius to the
utmost: for example, in trying to explain where he was at 2 o'clock some
night. . . . Whatever may be the cause of the abnormal condition in which
this distinguished scientist finds himself, we hope that it will soon be
removed, for we are certain that science in general, and Mr. Tesla in particular, will be all the richer when he gets married."

The absurd quidnunc who wrote this editorial would never, of
course, live to see his prophecy fulfilled. But neither would he be disappointed in Tesla's future scientific and technical achievements, for the
inventor was shortly to embark on one of the most extraordinary phases
of his altogether extraordinary career.

The event that signaled this new turn in Tesla's fortunes was another
long-distance telephone call from George Westinghouse. It was wonderful, astonishing news. The inventor quickly packed his bags and boarded
a train for Niagara Falls.

9. HIGH ROAD, LOW ROAD

It seemed almost too much success in such a short period. The Niagara Falls Commission, which for years had been swayed by the direful arguments of Edison and Lord Kelvin about the dangers of alternating current, announced in October 1893—just as Westinghouse had predicted—that it was awarding to his firm the contract to build the first two generators at Niagara.

The War of the Currents that had divided American industry so long and rancorously was to be settled with a victory for Tesla's system of AC and Westinghouse's perseverance.[1] No doubt this had resulted in large part from the unassailable visual testimony of their exhibitions at the Chicago World's Fair.

The war was to end with a compromise: General Electric was given the contract for building transmission and distribution lines from Niagara Falls to Buffalo. Both firms had submitted a proposal to install a Tesla polyphase generating system, for GE had obtained a license to use the Tesla patents and proposed to install a three-phase system. The Westinghouse proposal was for two-phase.

In 1895 the powerhouse was completed by Westinghouse and ready to deliver 15,000 horsepower of electricity, a truly phenomenal achievement for the times. The following year GE completed the transmission and distribution lines, enabling power to surge across twenty-six miles to run the lights and streetcars of Buffalo.

The harnessing of Niagara Falls proceeded on schedule. People spoke reverently of it as one of the official wonders of the world. Westinghouse built seven more generating units, which raised the production of electricity to 50,000 horsepower. General Electric constructed a second powerhouse that also used alternating current and built eleven more generators.

Another historic first soon followed. AC was delivered to one of its earliest and most significant customers, the Pittsburgh Reduction Company, which later became the Aluminum Company of America, or Alcoa.[2] The new metallurgical industry had been waiting for the high voltages that AC alone could supply. As Tesla had predicted, aluminum manufacture would soon permit the development of an aircraft industry.

An astounding aspect of the War of the Currents is that, like an ancient religious feud, it is still being waged. Anyone reading the national advertising campaign launched by General Electric in the late 1970s would have erroneously concluded that GE alone harnessed Niagara Falls and that Tesla was merely an also-ran among inventors.

Gardner H. Dales of the Niagara Mohawk Power Corporation, addressing the American Institute of Electrical Engineers (AIEE) on April 5, 1956, recollected more accurately:

"If there ever was a man who created so much and whose praises were sung so little—it was Nikola Tesla. It was his invention, the polyphase system, and its first use by the Niagara Falls Power Company that laid the foundation for the power system used in this country and throughout the entire world today. . . ."

Actually, however, Tesla's praises were well sung at this period and only later would it become convenient for the beneficiaries of his genius to grow forgetful. In the 1890s his name and achievements were almost constantly in headlines.

Newspapers and engineering journals alike saluted him. *The New York Times* declared that he owned the "undisputed honor" of making the Niagara enterprise possible, a sentiment echoed by George Forbes in *Electricity* (October 2, 1895). The achievement was covered widely in the world press. The Prince of Montenegro conferred upon him the Order of the Eagle. The coveted Elliott-Cresson Medal was awarded to him by the AIEE for his researches in high-frequency phenomena. And Lord Kelvin, now generous in his praise, declared that the inventor had "contributed more to electrical science than any man up to his time."

Soon alternating-current power systems were being built in New

York City for the elevated and street railways, for steam-railway electrification, and were even being extended to the Edison substations.

Nevertheless the inventor and Westinghouse continued to be torn and worried by sore losers. The company defended its alternating-current patents in some twenty court actions—including the one alluded to earlier that was determined by the U.S. Supreme Court—in each of which Westinghouse won a decisive victory. It filed actions against General Electric and others, and these too were successful. But as mentioned earlier, so much litigation created public confusion and left unhappy men. Some of these who had once praised Tesla now did their best to damage him.

B. A. Behrend, later vice-president of the AIEE and an acute observer of the contemporary scene, wrote: "It is a peculiar trait of ignorant men to go always from one extreme to another, and those who were once the blind admirers of Mr. Tesla, exalting him to an extent which can be likened only to the infatuated praise bestowed on victims of popular admiration, are now eagerly engaged in his derision."

Behrend found this deeply melancholy.

"I can never think of Nikola Tesla," he added, "without warming up to my subject and condemning the injustice and ingratitude which he has received alike at the hands of the public and of the engineering profession."[3]

Weary of the bickering and backbiting, the inventor returned to New York, more determined than ever to protect his time, aching to follow up half a dozen lines of research.

He began to achieve effects with high-voltage equipment that opened an infinity of possibilities. By learning to create artificial lightning he hoped not only to discover how to control the world's weather but also how to transmit energy without wires. And this in turn meshed with research that he hoped would enable him to build the first worldwide broadcasting system.

Gratifying results came when he achieved tensions of about one million volts using a conical coil. Instinctively he felt that instead of going to larger and larger apparatus for high voltages, he might accomplish the same thing with the proper design of a comparatively small and compact transformer.[4] This problem obsessed him, but not exclusively.

If some spectacular experiment seemed to defy the most elemental laws of electricity, Tesla cheerfully followed wherever it led. Sometimes it led in strange directions.

The radio tube, which involves the conduction of current through a vacuum, is, practically speaking, the original electronic device. Its accidental ancestor was a vacuum lamp invented by Edison in 1883. He was puzzled by what came to be known as the Edison Effect but saw no value in it; other scientists such as Sir William Preece, J. A. Fleming, Tesla, Elihu Thomson, and J. J. Thomson, however, were most interested. J. J. Thomson figured out that the observed phenomenon was caused by the emission of negative electricity, or electrons, passing from the hot element to the cold electrode. Edison, still puzzled and disappointed at not having found a good lamp, reported that the effect seemed to "impress some of the bulge-headed fraternity of the Savanic World." He himself moved on to more pressing concerns.

Tesla had begun developing vacuum tubes in the early 1890s, fully expecting them to be suitable for detecting the transmission of radio signals. Later he engaged a full-time glass blower and invented thousands of versions which he used both in radio research and for the production of light.

It was Fleming who, after studying the work of Edison and Preece, successfully applied the Edison Effect to the detection of radio signals, achieving increased sensitivity over the crystal detectors then used. In 1907 Lee De Forest would add the grid or control element to the Fleming diode, calling it the Audion, and the science of modern electronics would be fairly launched.

Yet long before this, Tesla was describing his work with vacuum bulbs and high-frequency currents, sharing his own fascination and puzzlement with his lecture audiences. Thus one day he placed a long glass tube, partially evacuated, within a longer copper tube with a closed end.

A slit was cut in the copper tube to disclose the glass within. When he connected the copper to a high-frequency terminal, he found the air in the inner tube brilliantly lighted although no current seemed to be flowing through the short-circuiting copper shell. The electricity, it seemed, preferred to flow through the glass by induction and pass through the low-pressure air rather than traversing the metal path of the outer tube.

In this the inventor saw a way of transmitting electric impulses, of any frequency in gases. "Could the frequency be brought high enough," he speculated, "then a queer system of distribution, which would be likely to interest gas companies, might be realized; metal pipes filled with gas—the metal being the insulator and the gas the conductor—supplying phosphorescent bulbs, or perhaps devices not yet invented."

In fact, what he was describing was the ancestor of the wave guide for microwave transmission.

Tesla was led by this line of exploration to one of his most grandiose conceptions, the "terrestrial night light"—a way of lighting the whole Earth and its surrounding atmosphere, as though it were but a single illumination. He theorized that the gases in the atmosphere at high altitudes were in the same condition as the air in his partially evacuated tubes and hence would serve as excellent conductors of high-frequency currents. The concept intrigued him for many years. He saw it as a means of making shipping lanes and airports safer at night, or as a way of illuminating whole cities without the use of street lights. One had only to transmit sufficient high-frequency currents in the right form to the upper air, at an altitude of 35,000 feet or even lower. When asked how he proposed to conduct his currents to the upper air, he merely replied that it did not present any practical difficulties. It was his habit never to disclose methods until he had tested them in practice, and this was one of his ideas that was to be put aside for lack of research capital.

Journalists continued to question him and to speculate. Some suggested that he planned to use one of his molecular bombardment tubes to project a powerful beam of ultraviolet rays into the atmosphere, ionizing the air through great distances and making it a good conductor of electricity of all kinds at high voltages. This, they theorized, would provide a conducting path to any desired height through which he could send high-frequency currents.[5] Later, when his great (and ill-fated) world-broadcasting tower was built on Long Island, the upper platform was designed to receive a bank of powerful ultraviolet lamps. Their purpose was never revealed.

At other times, Tesla talked of a plan for using both Earth and upper air as conductors of electricity and the stratum of air between as an insulator. This combination would form a kind of gigantic condenser, a means of storing and discharging electricity. If the Earth were electrically excited, the upper air would be charged by induction. The globe would be transformed into a Leyden jar, charging and discharging. A current flowing both in the ground and in the upper air would create a luminous upper stratum that would light the world. Was this how Tesla proposed to get his currents into the upper air? We do not know.

In his London lectures of 1892 he had lingered fondly over the description of a most peculiar and sensitive vacuum tube he had invented. Under the influence of a high-frequency current it would shoot off a ray that behaved with strange sensitivity to electrostatic and magnetic influences. With this tube he could make curious experiments.

When the bulb hung straight down from a wire and all objects were remote from it, Tesla could by approaching it cause the ray to fly to the opposite side of the bulb; and if he walked around the bulb the ray would always be on the opposite side of it. Sometimes the ray would begin to spin wildly around the bulb. With a small permanent magnet he could slow down or accelerate the spinning according to the position of the magnet. When most sensitive to the magnet, however, it was less sensitive to electrostatic influence. He could not make even the slightest motion such as stiffening the muscles of his hand, without causing visible reaction in the ray.

Tesla believed it was formed by an irregularity in the glass that prevented it from passing equally on all sides. Fascinated, he believed such a tool would be a valuable aid to investigating the nature of force fields.

"If there is any motion which is measurable going on in space," he said, "such a brush ought to reveal it. It is, so to speak, a beam of light, frictionless, devoid of inertia.

"I think that it may find practical applications in telegraphy. With such a brush it would be possible to send dispatches across the Atlantic, for instance, with any speed, since its sensitiveness may be so great that the slightest changes will affect it. If it were possible to make the stream more intense and very narrow, its deflections could be easily photographed."

He had closed his lecture with the comment: "The wonder is that, with the present state of knowledge and the experiences gained, no attempt is being made to disturb the electrostatic or magnetic condition of the Earth, and transmit, if nothing else, intelligence...."[6]

The little vacuum tube, however, was not to figure usefully in his plans as a detector of electrical disturbances or radio signals from a distance. It remained a curiosity item. When used by Tesla as a detector it was so difficult to adjust that it was unsuitable except for laboratory research.

But today, now that science has begun to take an interest in little-understood biological phenomena, Tesla's strange vacuum tube may hold new interest. It could, for example, have application in the control of autonomic functions of the body through biofeedback techniques. Or perhaps it might help us to understand the mysterious Kirlian effect. Kirlian photography, used in conjunction with high-frequency voltages of a Tesla coil, has created scientific interest in the human aura by disclosing to ordinary vision what may always have been apparent to psychics. Tesla's 1890s research showed that high-frequency currents move on or near the surface of conducting materials, similarly to the phenomenon of superconductivity. It has been speculated that coronas appearing in Kirlian photographs may be the modulation of some kind of "carrier field" surrounding life forms. (Acupuncture points also may be related to such force fields.) It is thus possible to entertain the suggestion of a contemporary electrical engineer that Tesla's hypersensitive vacuum tube might make an excellent detector not only of Kirlian auras but of other so-called paranormal phenomena, including the entities commonly called ghosts.

Since his return to New York, Tesla had lived almost a hermit's existence. Only on the most tempting social occasions were the inventor's friends any longer able to lure him from his laboratory. The late night fun and games had stopped. Robert and Katharine Johnson worried about him, warning him that all work and no play could bring on another break-down.

Katharine found the winter of 1893 passing slowly without his frequent company. In icy January she sent flowers to him in appreciation of some gesture. He found time to send her an article by Professor Crookes and a Crookes radiometer, a little heat-powered windmill that spun in an evacuated bulb, and which he considered (or said he did) "the most beautiful invention made." These small windmills, embodying in their simplicity Tesla's ideal of an elegant solution, may still be seen, in the windows of novelty shops, their blades silently "fanned" by the sun.

Although science was not her favorite subject, Katharine felt flattered and pleased. On a stormy afternoon in February she and Robert sat before an open fire, she feeling bored and restless. On the spur of the moment she wrote a note to Tesla and sent it off by messenger: "What are you doing these stormy days? We . . . are wondering if anybody is coming in this evening to cheer us up, say about 9, or at 7 for dinner. We are very dull and very very comfortable before an open fire, but two is too small a number. For congeniality there must be three, especially when it snows 'in my country.' Is that wonderful machine in order again and are you ready for the photographers and the thunderbolts and Juno and all the lesser gods and goddesses tomorrow? Come and tell us. We shall look for you, at 7 or 9."[7] But the machine was not in order, and the Johnsons were disappointed, Robert as much as Katharine.

Later in the spring of 1894, however, his experiments were far enough along for Tesla to invite Johnson, Joseph Jefferson, Marion

Crawford, and Twain to the laboratory to "take high-voltage sparks through their bodies" and to pose for the first photographs ever taken by gaseous tube lights.

Despite his absorption in science, it was characteristic of Tesla that he found time in May to write an article for Johnson's *Century* magazine on Zmaj Jovanovich, the chief Serbian poet. And the following spring he would be back in the pages of that journal with an article on his favorite hero, Luka Filipov.

Later on in the year he would give John Foord of *The New York Times* a major article (September 30, 1894) in which, in addition to describing his theory of light, matter, ether, and the universe, he claimed that 90 percent of the energy in electric lights was wasted and that in the future there would be no need to transmit power at all, not even wirelessly. "I expect to live to be able to set a machine in the middle of this room," he said, "and move it by no other agency than the energy of the medium in motion around us."

In this most productive period of his life, it is likely that he was at his happiest. No intimation of approaching disaster marred his days. He was still living restively at the staid Gerlach Hotel, and on its letterhead, in his most gracious style, he wrote to Katharine, accepting at last an invitation to dinner:

"Even dining at Delmonico's is too much of a high life for me and I fear that if I depart very often from my simple habits I shall come to grief. I had formed the firm resolve not to accept any invitations, however tempting; but in this moment I remember that the pleasure of your company will soon be denied to me (as I am unable to follow you to East Hampton where you intend *camping out* this summer)—an irresistible desire takes hold of me to become a participant of that dinner, a desire which no amount of reasoning and consciousness of impending peril can overcome. In the anticipation of the joys and of probable subsequent sorrows, I remain. . . ."[8]

In June 1894, a coy message came from Katharine in East Hampton, chiding him for ' "sending disappointing and cold-blooded telegrams to kind expectant friends." She added: "'In my country' one is *never* so cruel, especially after high honors when friends are longing to felicitate one. But on such occasions one is so genial and happy one cannot say no to a friend but must wish his friend as happy as himself. *This* is a friend 'in my country.'"[9] The honors to which she alluded were the LL.D. from Columbia College and the Order of St. Sava from the King of Serbia.

Shortly afterward she tried a variation on her usual routine, inviting Tesla and one of his gentleman friends to dinner. But he was firm (and perhaps wary), replying that he would attend provided there was a woman for every man, and said he would be pleased if she asked Miss Merington.

The summer passed and part of another winter with his friends almost never seeing him. He was intensely busy and apparently quite content, although perhaps sometimes during this period when his research seemed to lead in every direction at once, Tesla might have remembered with a smile Lord Rayleigh's well-meant advice about specializing.

Then, suddenly disaster struck. At 2:30 in the morning of March 13, 1895, his laboratory at 33–35 South Fifth Avenue caught fire. The six-story building in which it was located was destroyed, the cost to him being incalculable. All the expensive research apparatus that he and Kolman Czito had so laboriously built crashed right through from the fourth floor to the second where it came to rest, a mass of molten, reeking metal.[10]

Nothing was insured. But even had it been, it could not have covered his losses. Indeed, a million dollars, as he later said, could not have compensated for the resulting setbacks in his research. Stunned, sickened, he turned away from the ruins in the cold early morning and wandered through the streets in a trance, paying no attention to where he was or to the passing of time. The Johnsons frantically searched for him in his familiar haunts.

Newspapers all over the world reported the tragedy: "Work of half a lifetime gone." "Fruits of Genius Swept Away." In London the *Electrical*

World reported that the greatest loss was the physical collapse of the inventor.[11] Charles A. Dana of the New York *Sun* paid him the highest tribute: "The destruction of Nikola Tesla's workshop, with its wonderful contents, is something more than a private calamity. It is a misfortune to the whole world. It is not in any degree an exaggeration to say that the men living at this time who are more important to the human race than this young gentleman can be counted on the fingers of one hand; perhaps on the thumb of one hand."[12]

Only his closest assistants knew the dazzling scope of his advanced researches in radio, wireless transmission of energy, and guided vehicles, or that he was achieving effects with what the world would soon know as X rays, and also nearing a breakthrough in the potentially lucrative industrial discovery of a means of producing liquid oxygen. It may have been the latter volatile substance that caused the blaze—apparently started from a gas jet on the first floor near oil-soaked rags—to explode so rapidly through the entire building.

An emotional letter from Katharine, written the day after the fire, finally reached him. She told of their search and the hope of consoling him in his "irreparable loss."

"It seemed as if you too must have dissipated into thin air. . . . Do let us see you again in the flesh that this awful thought may vanish," she implored. "Today with the deepening realization of the meaning of this disaster and consequently with increasing anxiety for you, my dear friend, I am even poorer except in tears, and they cannot be sent in letters. Why will you not come to us now—perhaps we might help you, we have so much to give in sympathy. . . ."[13]

The degree to which this strangely unresponsive man had begun to affect her life and happiness was no longer a question in her mind.

10. AN ERROR OF JUDGMENT

At this crucial point in his life Tesla, for all his worldwide fame, was close to being broke. The destroyed laboratory of the Tesla Electric Company was owned in part by A. K. Brown and another associate. There were no longer any royalties from his alternating-current patents in America, nor any salary from Westinghouse. He had invested everything he owned in equipment for research. His only current resources were royalties from German patents on his polyphase motors and dynamos, which were a drop in the bucket compared to what he would need to rebuild and refurnish a laboratory.

He was not downcast for long, however, consoling himself with the fact that his ongoing research was still vivid in his mind and that the loss was merely a setback.

To the rescue came Edward Dean Adams, the financier who had organized the International Niagara Commission when competing technologies were being examined for the harnessing of Niagara Falls. He was also president of the Morgan-backed Cataract Construction Company which, holding the charter for development of power at the Falls, had chosen Tesla's polyphase system. Hence he was well acquainted with the inventor's record and impressed by his genius.

Adams proposed not only to form a new company for his continued research, with a capitalization of $500,000, but he himself offered to subscribe to $100,000 in stock. For starters he gave Tesla $40,000.

The inventor at once began combing New York City for a new laboratory and soon found a location at 46 East Houston Street. He had a telephone installed (Spring 299) and began firing off a barrage of oral and written SOS's to Westinghouse for replacement machinery.

To Albert Schmid, general superintendent of the Pittsburgh head-quarters, he wrote: "You will greatly oblige me if you will do what is in your power to ship what is required with the least possible delay." And again: "Let me know immediately . . . what is the smallest size rotating two-phase transformer you have in stock. . . ."[1]

Only days later he asked that the machinery be sent by costly express rather than as freight, being in an agony to get on with inter-rupted research, especially in wireless, or radio, where the interna-tional race had already begun.

Both Edison and William H. Preece, head of the British Postal Telegraph System, had been working with primitive "wireless" that used an inductive effect. That is to say, Edison had sent a message from a mov-ing train via a telegraph wire strung on poles along the track, bridging the intervening feet by induction. But such systems could work only over short distances, and Edison had characteristically lost interest.

More to the point, Sir Oliver Lodge just the previous year had trans-mitted Morse signals between two buildings at Oxford University, a dis-tance of several hundred feet. He had built a transmitter and receiver by putting a Hertz spark gap in a copper cylinder open at one end, thereby producing a beam of ultra-short-wave oscillations.

Tesla explained to the Westinghouse superintendent that the machinery he was ordering was to be used in connection with his oscillators and a high efficiency was important. "Please," he pleaded, "do not spare any pains of expense. I shall rely as to the price entirely on the fairness of the Westinghouse Company. I believe that there are gentlemen in that company who believe in a hereafter."[2]

Assurances came from the vice-president and general manager that the equipment was being shipped and that the price would be as low as possible. After all, as Tesla occasionally reminded them, they benefited from valuable promotion when he used their equipment for his demonstrations.

To Schmid he wrote again, exhorting him to make the rotary trans-former excellent in every way. To C. F. Scott, chief electrician at

Pittsburgh, he urged that the schedule for building the transformer be advanced: "My work has been suddenly interrupted just as I was at the most interesting stage of the development of certain ideas, and I need very much my apparatus to begin work anew."

Only weeks later Scott received another message in the same vein of urgency: "This kind of work is almost essential to my health, and I hope that its resumption will have a good effect upon me."

Even while buying equipment Tesla was mulling over the tempting offer of Edward Dean Adams to join forces in a new company that would mean the powerful financial backing of the House of Morgan. But he was very leery of it, having seen the Morgan takeover of both the Thomson-Houston Company and the Edison Electric Company to form General Electric. And he remembered well how they had coveted and threatened the autonomy of Westinghouse. So he made one of his many errors of judgment in finance, accepting the $40,000 from Adams but rejecting the larger alliance.

His good friend Johnson was only one of those who thought him mistaken for cutting himself off from the security represented by the House of Morgan. Tesla sighed, spread his long hands expressively, and spoke of protecting his precious freedom. Undoubtedly he believed that with the $40,000 he could bring to commercial stage at least some of the inventions on which he was currently nearing success. As usual, however, he underestimated the time and the costs involved.

"No other discovery within my lifetime," Michael Pupin wrote, "had ever aroused the interest of the world as did the discovery of the X rays. Every physicist dropped his own research problems and rushed headlong into the research. . . ."[3]

Roentgen announced his discovery in December 1895. Edison, mired in a perennial and ultimately disastrous effort to mine ores magnetically, quickly sent a wire to a former associate, urging him to drop everything and join a group to experiment on "Rotgens" (sic) new radiations. "We could do a lot before others get their second wind," he said.

The opportunity to see the internal structure of the human body captivated everyone, and it was obvious to scientists and engineers that some sort of fluoroscopic screen would be needed to register the rays after their passage through the body.

The ways in which Edison, Pupin, and Tesla severally proceeded with their X-ray research were characteristic of their different personalities.[4] Edison, seeing where the commercial potential lay, began at once to test various chemicals and quickly reported that calcium tungstate crystals gave a good fluorescence on a screen. Then he rushed to the Patent Office.

Pupin noted in his diary that American physicists had paid little attention to vacuum-tube discharges and that, to the best of his knowledge, *he* was the only American physicist with any experience. Hence when Roentgen's discovery had been announced, "I was, it seems, better prepared than anybody else in this country to repeat his experiments and succeeded, therefore, sooner than anybody else on this side of the Atlantic."[5] He claimed to have obtained the first X ray in the United States on January 2, 1896, two weeks after the discovery was announced by Roentgen in Germany.

This was curious in view of Tesla's pioneer love affair with vacuum tubes, which he had demonstrated in his series of lectures in 1891, 1892, and 1893. Although Tesla always gave full priority to Roentgen, he had spoken then of both "visible and invisible" rays when demonstrating his molecular-bombardment lamp and other gaseous lamps, and he was using uranium glass and a variety of phosphorescent and fluorescent substances for detecting radiation. During experiments he carried on in the fall of 1894 with the assistance of the Manhattan photographers Tonnelé & Company on the radiant power of phosphorescent bodies, "a great number of plates showed curious marks and defects." It was just as he was beginning to explore the nature of these phenomena that his laboratory burned down.[6]

When Professor Roentgen announced his discovery of X rays in December of that year, Tesla immediately forwarded shadowgraph pictures to the German, who replied: "The pictures are very interesting. If you would only be so kind as to disclose the manner in which you obtained them."

The Pupin claim to have been the first in the United States experimenting with vacuum-tube discharges would have been unlikely even if Tesla had not preceded him. Apparently they were being investigated in numerous laboratories in America and Europe, and after Roentgen's announcement a dozen claims were made to "firsts" in X ray. Tesla never made any such claim in his own behalf. The first clinical radiograph in North America is said by some to have been made in the basement of Reid Hall at Dartmouth College on February 4, 1896,[7] by a laboratory assistant.

But Edward R. Hewitt, an inventor doing photographic research at this time, has left an intriguing anecdote. His own researches "began on the morning when Nikola Tesla took a picture of Mark Twain under a Geissler tube which proved to be no picture of Twain but a good one of the adjusting screw of the camera lens."

"Neither Tesla nor Hewitt," wrote Noel F. Busch in *Life* magazine (July 15, 1946), "realized until a few weeks later, when Roentgen announced the discovery of X rays, that the picture of Twain was in fact an example of X-ray photography, the first ever made in the U.S." This is, of course, hardly proof of priority of invention, which includes much more than achieving accidental effects, but it does suggest how far advanced Tesla's research was at this time.

Whereas Edison hastened to try to profit from Roentgen's discovery and Pupin was quick to try to share in its glory, Tesla's less self-interested response was to begin an exhaustive series of experiments in X-ray phenomena and technique, the results of which he published, beginning in March 1896, in a series of articles in the *Electrical Review*.[8]

While his competitors were using Roentgen tubes for the production of weak shadows of hands and feet, Tesla claimed to be making forty-minute photographs through the human skull at forty feet. If this were true, he would have to have been using equipment far more advanced than anything we now believe existed at that time.

On April 6, 1896, Professor Pupin reported to the New York Academy of Sciences: "Every substance when subjected to the action of the X rays becomes a radiator of these rays," and thus claimed to have discovered secondary radiation. But Tesla had already publicly reported in the *Electrical Review* (March 18, 1896): "I have lately obtained shadows by *reflected rays only*," and described how he had excluded direct rays to obtain this effect. In testing various kinds of metals, he discovered that the most electropositive made the best "reflectors" of Roentgen rays.

Many competitors had now entered the field, including such well-known inventors as A. E. Kennelly and Edwin J. Houston, who used a simple form of Tesla coil to produce Roentgen rays. The practical Edison, delighted by the public's enthusiasm, made a number of fluoroscopes in the form of boxes with peepholes and placed them on display at the Electrical Exposition of 1896 at the Grand Central Palace in New York. This was the first opportunity Americans had to see their skeletal shadows, and they clamored for a place in line. Many were disappointed at not being allowed to view their brains in action. A gambler wrote to Edison asking for an X-ray device with which he could play against the faro bank.[9]

Prudes worried about the danger of unscrupulous manufacturers making X-ray binoculars, enabling voyeurs to strip them naked as they strolled in Sunday finery along Fifth Avenue. Well into the 1940s the foot X-ray machine in shoe stores would provide a consumer come-on for small-town America.

On the theory that blindness might be cured with X rays, numerous "treatments" were given by doctors. To the contrary, as is now known, radiation can cause "flashes" in the eye and, with overexposure, cataracts. Tesla pointed out that no evidence whatever existed for the blindness "cure" and discouraged the building up of false hopes as cruel. Edison too deplored this yet as a recent biographer notes, "he jumped in and conducted the tests along with other scientific men and doctors."[10]

Tesla's research, which was fundamental and well-documented, con-

vinced him that X rays were composed of discrete particles. This proved
to be incorrect; but so were almost everyone else's theories in this early
period. Dr. Lauriston S. Taylor, a radiological physics consultant and
recent past president of the NCRP, says, "Nevertheless his reasoning was
good and much to his credit."

Almost simultaneously, at Cambridge University in England, the
physicist Joseph J. Thomson had built a vacuum tube with two charged
plates and a fluorescent screen. He discovered that the radiation caused
by the flow of currents made dots on the screen. Both magnetic and electric fields deflected the rays of electricity, which convinced him that they
were charged particles. Since the ratio of the charge on the particles to
their mass was always the same, he hypothesized that he had discovered
"matter in a new state" from which all the chemical elements were built
up. Some years later Thomson was credited with having discovered (in
1897) the *electron*—a very light particle associated with the elementary
charge of negative electricity and the fundamental building block of the
atom.

Max Planck in 1900 proposed a law for electromagnetic radiation—
the quantum theory. And five years later Einstein explained, with his
special theory of relativity, that all radiation, though it consisted of
quanta of different amounts of energy, traveled at the speed of light. His
fundamental equations described the exchange of energy that took place
when radiation and matter interacted.

From this new realm of physics came knowledge of the properties of
different kinds of electromagnetic radiations. Radio waves, at the lowest
frequency, stretched for thousands of miles. In the order of rising frequencies came microwaves, infrared radiation, visible light, ultraviolet
radiation, X rays, and gamma rays—the latter incredibly short.

Tesla and other early experimenters with X rays were exploring
treacherous territory. It was clear that radiation would be useful in
detecting foreign objects in the body or bone fractures, but its full medical potential and the effect of such rays on human health entailed dangerous trial-and-error research.

"Yet in spite of some grievous accidents with X rays for their first twenty-five years," says Dr. Taylor, "there were surprisingly few who suffered from overexposure—certainly not everyone."[11]

Tesla, entranced with the novel and mysterious force, was one of those who at first refused to believe there was danger. Convinced he had discovered a way of "stimulating" his brain, he exposed his head repeatedly to radiation.

"An outline of the skull is easily obtained with an exposure of 20 to 40 minutes," he wrote. "In one instance an exposure of 40 minutes gave clearly not only the outline, but the cavity of the eye . . . the lower jaw and connections to the upper one, the vertebral column and connections to the skull, the flesh and even the hair."[12]

He noted strange effects: " . . . a tendency to sleep and the time seems to pass away quickly. There is a general soothing effect and I have felt a sensation of warmth in the upper part of the head. An assistant independently confirmed the tendency to sleep and a quick lapse of time."

From such effects he was more than ever inclined to believe that the radiation was of material streams penetrating the cranium. And he was first to suggest that X rays would be used therapeutically—perhaps to "project chemicals into the human body."[13]

It is difficult to gauge at this date the degree of exposure to which he subjected himself. And indeed, insofar as the brain is concerned, it is still not known what its physiological tolerances are to high-energy radiofrequency fields.

Edison damaged his eyes with X-ray exposure. One of his assistants contracted a gradually spreading skin cancer from which he died several years later.

Tesla described carefully the effects of X rays upon his own eyes, body, hands, and brain, differentiating between skin burns and what he considered to be internal effects. In the spring of 1897 he was mysteriously ill for several weeks. He reported receiving frequent sudden and painful shocks in the eye from X-ray equipment. His hands were repeatedly exposed.

"In a severe case," he wrote, "the skin gets deeply colored and black-ened in places, and ugly, ill-foreboding blisters form; thick layers come off, exposing the raw flesh. . . . Burning pain, feverishness and such symptoms are of course but natural accompaniments. One single injury of this kind in the abdominal region to a dear and zealous assistant—the only accident that ever happened to anyone but myself in all my labora-tory experience—I had the misfortune to witness."[14]

This had followed an exposure of five minutes, only a few inches from a highly charged tube. But apart from skin damage, he noted that such radiation caused a feeling of warmth deep in the flesh, a fact that was to inspire his continuing work in therapeutics.

It is now known that X rays may be of two kinds—"hard" or "soft," meaning that the latter have longer wavelengths and lower energies. They are more easily absorbed than hard X rays. Even so they are of high energy compared to ultraviolet or visible light rays.

Tesla's research very quickly convinced him that safety measures were needed. He lectured to the New York Academy of Science on April 6, 1897, on the practical construction and safe operation of X-ray equip-ment as well as reporting his observations of the dangers of Roentgen rays. He had already experimented with various metal protective devices, and soon thereafter lead shields came into general use.

An important figure entered the inventor's life at this juncture. In prepar-ing for his Academy of Science lecture he was supplied with lantern slides and cathode tubes by an eager new assistant named George Scherff.

At first his secretary, Scherff was to become a financial and legal adviser, bookkeeper, office manager, stockholder, factotum, friend, and during acute financial squeezes, a nearly-always-reliable source of small loans. Devoted through good times and lean, he was to become Tesla's most loyal and least dispensable employee.

Scherff never complained about long hours, scanty rewards, or the occasional thoughtlessness of his boss. If it meant depriving his own family to help Tesla out of a tight spot, the good and frugal Scherff would manage. He never questioned the fact that he was always Mr. Scherff, the loyal functionary, never an intimate or social equal. He truly worshipped Tesla, learned more about his affairs than anyone else, and would go to his grave with sealed lips where the inventor's private matters were concerned. If ever there was a faithful friend standing behind a great man, it was George Scherff behind Nikola Tesla.

Many people continued to worry about why no good woman could be seen standing behind the celebrated inventor. Important people were expected to procreate for the good of the country. Urging Tesla to get married in 1896 were not just gossip columnists. Technical journals like the *Electrical Review* of London, the *American Electrician,* and the *Electrical Journal* also took up the hue and cry.

Tesla's expertise in handling such queries is apparent at the end of a long interview he gave to a reporter for the New York *Herald,* who came upon him one night slumped in a café at a late hour, looking haggard and tired. He was still brooding at times over the setbacks he had suffered when his laboratory burned, but it was apparent to the reporter from his pallor and the look in his eyes that something was seriously troubling him.

"I am afraid," began Tesla, "that you won't find me a pleasant companion tonight. The fact is, I was almost killed today." [15]

He had gotten a shock of about 3.5 million volts from one of his machines.

"The spark jumped three feet through the air," he said, "and struck me here on the right shoulder. I tell you it made me feel dizzy. If my assistant had not turned off the current instantly it might have been the end of me. As it was, I have to show for it a queer mark on my right breast where the current struck in and a burned heel in one of my socks where it left my body. Of course the volume of current was exceedingly small, otherwise it must have been fatal." [16]

It is possible that he was even minimizing the accident because of Edison's long campaign against "deadly AC."

The reporter asked how far sparks could travel.

"I have frequently had sparks from my high-tension machines jump the width or length of my laboratory, say thirty to forty feet," he said. "Indeed, there is no limit to their lengths, although you can't see them except for the first yard or so, the flash is so quick. . . . Yes, I am quite sure I could make a spark a mile long, and I don't know that it would cost so much either."

Asked whether he had suffered many accidents while working with electricity, he said, "Very few. I don't suppose I average more than one a year, and no one has ever been killed by one of my machines. I always build my machines so that whatever happens it cannot kill anyone. The burning of my laboratory two years ago was the most serious accident I ever had. No one knows what I lost by that."

For a moment he sat reflecting. Then, speaking in the third person, he began to explain the main source of sadness in a prolific inventor's life.

"So many ideas go chasing through his brain that he can only seize a few of them as they fly, and of these he can only find the time and strength to bring a few to perfection. And it happens many times that another inventor who has conceived the same ideas anticipates him in carrying one out of them. Ah, I tell you, that makes a fellow's heart ache."

When the laboratory burned, he said, there was destroyed with it the apparatus he had devised for liquefying air by a new method. "I was on the eve of success, and in the months of delay that ensued, a German scientist solved the problem. . . ."

It was Linde who anticipated him in this important commercial breakthrough of liquid oxygen. Tesla had been seeking a means of refrigeration for the artificial insulation of electrical mains.

"I was so blue and discouraged in those days," he said, "that I don't believe I could have borne up but for the regular electric treatment which I administered to myself. You see, electricity puts into the tired body just what it most needs—life force, nerve force. It's a great doctor, I can tell you, perhaps the greatest of all doctors."

Asked if he were often depressed, he said, "Perhaps not often. . . . Every man of artistic temperament has relapsed from the great enthusiasms that buoy him up and sweep him forward. In the main my life is very happy, happier than any life I can conceive of."

He described the overmastering excitement of his research. "I do not think there is any thrill that can go through the human heart like that felt by the inventor as he sees some creation of the brain unfolding to success. . . . Such emotions make a man forget food, sleep, friends, love, everything."

It was as if he had purposely led the reporter to the next question. Did he believe in marriage "for persons of artistic temperament"?

Tesla considered carefully.

"For an artist, yes; for a musician, yes; for a writer, yes; but for an inventor, no. The first three must gain inspiration from a woman's influence and be led by their love to finer achievement, but an inventor has so intense a nature with so much in it of wild, passionate quality, that in giving himself to a woman he might love, he would give everything, and so take everything from his chosen field. I do not think you can name many great inventions that have been made by married men."

Whether this struck the interviewer as a sly put-down of Edison, with his two marriages, he did not indicate.

Tesla hesitated and then, adverting to his single estate, added with what the reporter described as pathos, "It's a pity too, for sometimes we feel so lonely."

11. TO MARS

Letters from Katharine betrayed both the mercurial state of her emotions and the steady state of her interest in Tesla. At this remove of time, it is difficult to know what to make of these curious missives. Effusive and intimate, they sometimes seem to stop just short of becoming love letters, but if Katharine was tending in that direction, Tesla gave her little encouragement.

On April 3, 1896, she invited him to their home, commenting that although he had looked ill when she had last seen him, yet he had managed to cheer her up, and "now I need to be brought up again." She mentioned that it was Easter. "I have always wondered when the great changes are in progress if you know of them," she wrote. "Do you know when Spring is near? It used to make me so happy and now it brings me only sorrow. It means so much that I would fain escape . . . disintegration, separation. I wish that I, like you, could go on forever and forever in the same routine, without break, living my own life, as you say you do. I do not know whose life I live, it has not seemed my own. You must come tomorrow evening, you see."[1]

The Johnsons spent a part of that summer in Maine, but separation from the inventor only increased Katharine's sadness and her concern for his health.

"You are making a mistake, my dear friend, almost a fatal one," she wrote. "You think you do not need change and rest. You are so tired you do not know what you need. . . ."[2]

In reply to these warm letters, Tesla alternately teased her or sent flowers when he thought of it. Perhaps he sensed he might be on treacherous ground. Robert was also his friend, and Robert loved Katharine, and. . . . But at least he probably did not have to worry about his own feelings. He had scarcely ever known a moment of vulnerability.

With Johnson he exchanged notes on religion, poetry, and whether or not he should pose for a certain painter of fashionable portraits for the May number of *Century*. A casual effusiveness had entered their correspondence, a "Dear Luka" from Tesla to Johnson saying, "I am glad to know that you still love me. . . ."[3]

Although he was no orthodox believer, Tesla commended religion as an excellent thing for others. In this period when anxiety over his inventions was stretched almost beyond bearing, and his pocketbook was equally thin, he became interested in Buddhism. It and Christianity, he believed, were to be the most important religions of the future. He therefore sent a book on Buddhism to Johnson, who replied: "Sir Knight: I did not know you were enlisted on that side of the campaign, but now when I read it I shall think of you even more frequently than usual—which is by no means seldom, let me assure you."[4]

Days later when the Johnsons again invited him to dinner, he joked about his weakness for elegant people: "If you have visitors (ordinary mortals) I will not come. If you have Paderewski, Roentgen, or Mrs. Anthony—I *will* come. Please answer."[5]

Christmas of that year was not a happy occasion for Katharine, despite or perhaps because of the usual efforts at family gaiety. She felt trapped. Although her children and husband were dear to her and she usually enjoyed the social round, a vital part of her life seemed missing. Was it worth living only for the slow disintegration she felt?

The day following Christmas she wrote to Tesla:

"I have tried several times to thank you for the roses. They are before me as I write—so strong, so superb in color. . . . I must always when I write to you make several attempts, a system of repression because I can never express what I would say. I did not mean to be severe the other evening. I was only wrapped up in disappointment. I miss you very much and wonder if it is always to go on this way and if I can ever become accustomed to not seeing you. However I am glad to know that you are well and happy and prosperous. With every kind wish for the New Year my dear friend."[6]

Typically, when Tesla got around to responding, he tried to lighten the mood with chiding. He only succeeded in being cruel, going on about how he had found her sister, whom he had recently met, much more pretty and charming than she. Then he went back to work.

After the lectures of 1893 in which he had described in detail the six basic requirements of radio transmission and reception, he had built equipment that could be operated between his laboratory and various points within New York City. The fire had destroyed all this and had set back his research, but by the spring of 1897, with financial help from Adams and strong support from Westinghouse, he was prepared to move ahead.

He announced to the *Electrical Review* in August, before filing his basic radio patents, that successful tests had been made, but the report was guarded and general: "Already he has constructed both a transmitting apparatus and an electrical receiver which at distant points is sensitive to the signals of the transmitter, regardless of earth currents or points of the compass. And this has been done with a surprisingly small expenditure of energy."

By disturbing the "electrostatic equilibrium" at any point on the Earth, the *Review* explained, the disturbance could be distinguished at a distant point and thus "the means of signalling and reading signals becomes practicable once the concrete instruments are available." By actual testing, said the report, he "has really accomplished wireless communication over reasonably long distances . . . and has only to perfect apparatus to go to any extent. . . ."[7]

Tesla made tests from a boat chugging up the Hudson River, carrying the receiving set twenty-five miles from his new laboratory on Houston Street. And this was only a fraction of what his instruments were capable of doing.

He filed his basic patent applications No. 645,576 and 649,621 on

September 2, 1897, and they were granted in 1900. Later, as we have noted, they would be contested in long litigation by Marconi; but first Tesla would sue the Italian for infringement.[8]

In 1898 he filed and was granted patent No. 613,809 which described radio remote control for use in guided vehicles. Here was yet another potentially spectacular application of wireless transmission. He could scarcely wait to show the public not just radio *or* the first breakthrough in automation, but both at once.

The year before when speaking at Buffalo on the occasion of introducing Niagara Falls power, GE having just completed its lines, Tesla had declared that he now hoped to see the fulfillment of his fondest dream, "namely, the transmission of power from station to station without the employment of any connecting wire. . . . "[9] The visiting dignitaries—engineers, industrialists, financiers—had listened with mixed emotions. This gifted madman seemed bent on making whole systems obsolete as soon as they came into being, and just when they promised to start earning profits. But soon newspapers around the world were announcing that he had developed equipment that not only would transmit energy and intelligence *through the Earth* for a distance of twenty miles, but that he also could send it wirelessly through the air.[10]

And so certain was Tesla that he now claimed that communication with Mars would be possible in a short time.

An announcement was carried by the *Electrical Review* describing how Mr. Tesla had invented apparatus "capable of generating electrical pressures vastly in excess of any heretofore used," with which the current "can be conducted to a terminal maintained at an elevation where the rarefied atmosphere is capable of conducting freely the particular current produced. At a distant point where the energy is to be used commercially a second terminal is maintained at about the same elevation to attract and receive the current, and to convey it to earth through special means for transforming and utilizing it."[11]

The article was illustrated with streamers representing electrical pressure of 2.5 million volts pouring from a single coil. Other publications showed huge stationary balloons being used to maintain the terminals at required elevations.

"Tesla now proposes," the *Electrical Review* continued, "to transmit without the use of any wires through the natural media—the earth and the air—great amounts of power to distances of thousands of miles. This will appear a dream, a tale from the 'Arabian Nights.' But the extraordinary discoveries Tesla has made during a number of years of incessant labor ... make it evident that his work in this field has passed the stage of laboratory experiment, and is ready for a practical test on an industrial scale. The success of his efforts means that power from such sources as Niagara will become available in any part of the world regardless of distance." [12]

Some of the articles appearing at this time reported the goal as a *fait accompli*, carrying such headlines as, "Tesla Electrifies the Whole Earth." Michael Pupin read Tesla's claim about being able to communicate with Mars and uttered a mute appeal to the patron saint of transplanted Serbs. Along with other scientific colleagues, he wondered, What next? Long ago as a boy herding cattle along the military frontier of Serbia, he had learned about the importance of the Earth as a conductor of acoustical resonance. He and the other boys had stuck their knives into the earth at night, falling asleep with their ears against the blades. The merest sound of moving cattle or of marauding Romanians stealing through the cornstalks would quickly awaken them.

Later Pupin realized that an oscillator sending out electrical waves would penetrate longer distances when one of its sides was connected to the Earth. But to speak of sending wireless signals to Mars seemed palpable nonsense, "because there would not be the acoustical resonance of earth to cover great distances."

Such minor considerations did not deter Tesla, however, as he built equipment that exceeded anything ever designed before. He built many shapes, sizes, and varieties of Tesla coils, or high-frequency transformers, including a flat-spiral resonant transformer that represented a beautiful evolution in design and with which he could produce electromotive forces of many millions of volts.

One of the major problems associated with very high-voltage apparatus is the loss due to corona and other spurious discharges, which severely "drag down" the output and ultimately limit maximum capability. To these problems Tesla succeeded in evolving elegant solutions.

He considered the ultimate design to be a transformer having a secondary in which the parts, charged to a high potential, were of considerable area and arranged in space along ideal enveloping surfaces of very large radii of curvature, and at proper distances from one another, thereby insuring a small electric surface density everywhere. Thus no leak could occur even if the conductor were bare. This design was exemplified in his flat-spiral coil.

In his laboratory he had installed a two-turn primary circuit running all around the large room and it was this coil, plus the associated circuit interrupters, that he would later ship to Colorado to drive his magnifying transmitter. The primary was buried in the ground, and it probably had such special characteristics as a very large diameter and multistrands.

With such equipment, he felt, there were no limits: a message could be sent to Mars almost as easily as to Chicago. "I found that there was practically no limit to the tension available," he wrote in the *Electrical Review*, and "I discovered the most important of all facts arrived at in the course of my investigation in these fields. One of these was that the atmospheric air, though ordinarily a perfect insulator, conducted freely the currents of immense electro-motive force producible by such coils. . . . So great is the conductivity of the air, that the discharge issuing from a single terminal behaves as if the atmosphere were rarefied. Another fact is that this conductivity increases very rapidly with the rarefaction of the atmosphere and augmentation of the electrical pressures, to such an extent that at barometric pressures which permit of no transit of ordinary currents, those generated by such a coil pass with great freedom through the air as through a copper wire."[13]

He had proved conclusively, he said, that great amounts of electrical energy could be transmitted through the upper air strata to almost any distance. And he learned what he considered an equally important fact: that the discharges of an electromotive force of a few million volts excited powerful affinities in the atmospheric nitrogen, causing it to combine with oxygen and other elements. "So energetic are these actions and so strangely do such powerful discharges behave," he said, "that I have often experienced a fear that the atmosphere might be ignited, a terrible possibility, which Sir William Crookes, with his piercing intellect, has already considered. Who knows but such a calamity is possible?"

Electrical resonance was not Tesla's original idea, for Lord Kelvin had introduced the mathematical potential of the condenser discharge; but Tesla exhumed the equation and gave it vibrant life.

In the 1899 *Electrical Review* article in which Tesla expressed fear of setting fire to the sky, several startling photographs appeared of the inventor working with the apparatus he had been building.[14] One records a spectacular display of lightning achieved with pressure of about eight million volts in an experiment for transmitting electrical energy great distances without wires. Another shows the inventor holding a disconnected, brilliantly lighted vacuum bulb of 1,500 candlepower, the light being used for the photograph. The frequency is measured in millions per second.

A third shows Tesla in brilliant relief, with a coil energized by the waves of a distant oscillator and adjusted to the capacity of his own body, which is preserved from injury "by maintaining a position at the nodal point, where the intense vibration is little felt." The pressure on the end of the coil, which is illuminated by powerful streamers, is nearly half a million volts.

A final photograph in this eerily remarkable series bears the caption: "In this experiment the operator's body is charged to a great pressure by a direct connection with an oscillator. The photograph shows a conducting bar, carrying on the end a sheet of tin of determined size, held in hand. The operator is on the top of a stationary electrical wave and the bar and sheet are both illuminated by the violently agitated air surrounding them. One of the vacuum tubes used in lighting the laboratory, though at considerable distance on the ceiling, glows brightly, being affected by the vibrations transmitted to it from the operator's body."

Tesla delighted in such magic, but for critics who might think him more interested in effects than utility, he added that there were to be mundane rewards as well. With the tools of electrical resonance and circuits in exact synchronism, he said, nitrogen could be extracted from the air and valuable fertilizer manufactured. Also light, "diffusive like that of the sun," could be produced with an economy greater than that obtainable in the usual ways and with lamps that never burned out.

His dreams were Utopian: Earth delivered from hunger and toil; easy world communication; control of weather; a bountiful supply of energy; limitless light; and last but not least, a link with the forms of life he was convinced existed on other planets. Martians he regarded as a "statistical certainty."

Meanwhile, for his friends of a more pedestrian nature, life continued as usual. Katharine sent him a poignant and critical letter, inviting him to yet another party and reminding him that he was neglecting his friends. The Johnson children were growing up, and she could foresee a day when even they would have no need of her. Time raced, and she was suffering from intimations of mortality: "Do leave aside the millionaires, high-sounding titles, the Waldorf and Fifth Avenue . . . ," she wrote, "for some simple everyday people who are distinguished only by a great weakness. . . .

"I have heard lots of things about you—I am sure some of them you don't know of yourself, and I am just dying to tell them all to you, but of course you would not care to hear them. Do you know that I am going abroad in the Spring, the early Spring, and who knows, perhaps these familiar scenes may know me no more. So if you have not forgotten me entirely, or forgotten to be fond of me—*I have forgotten to forget.* You had better come now and again.

" 'O how fast the days are flitting.' There are so few days left in my years, now it is Autumn and we are returning from exile, and then it is Spring and we are taking it up again, the interminable summer begins, there is no winter. Be human, be kind and come. You know it is Robert's party. Perhaps you will come for him." [15]

He emerged from his laboratory and went to the party. For a time he tried to be more thoughtful. In a note to the "Palais Johnson" he mentioned Luka's "great translations of Serbian poetry," and said he had sent three copies of his book to "three queens—American queens, I might add." He invited the Johnsons to a celebration at the Waldorf—"before I run out of money." And he sent a frivolous note to "Mrs. Johnston, the Belle of the Ball," of which, many years later, Agnes Johnson Holden was to write on the envelope: "Joke played on mother by Mr. Tesla, disguising handwriting and misspelling her name."

With partying resumed it was for a while almost like old times. But soon the seduction of the laboratory claimed him again. Tesla for a long time had been exploring the area of mechanical vibrations—as for example with the platform on which he had allowed Mark Twain to experiment for fun and health. Almost at once he had begun to produce unexpected effects.

One day in 1898 while testing a tiny electromechanical oscillator, he attached it with innocent intent to an iron pillar that went down through the center of his loft building at 46 East Houston Street, to the sandy floor of the basement.

Flipping on the switch, he settled into a straight-backed chair to watch and make notes of everything that happened. Such machines always fascinated him because, as the tempo built higher and higher, they would establish resonance with first one object in his workshop and then another. For example, a piece of equipment or furniture would suddenly begin to shimmy and dance. As he stepped up the frequency, it would halt but another more in tune would take up the frantic jig and, later on, yet another.

What Tesla was unaware of on this occasion was that vibrations from the oscillator, traveling down the iron pillar with escalating force, were being carried through the substructure of Manhattan in all directions. (Normally earthquakes are more severe at a distance from their epicenter.) Buildings began to shake, windows shattered, and citizens poured onto the streets in the nearby Italian and Chinese neighborhoods.

At Police Headquarters on Mulberry Street, where Tesla was

already regarded with suspicion, it soon became apparent that no other part of the city was having an earthquake. Two officers were dispatched posthaste to check on the mad inventor. The latter, unaware of the shambles occurring all around his building, had just begun to sense an ominous vibration in the floor and walls. Knowing that he must quickly put a stop to it, he seized a sledgehammer and smashed the little oscillator in a single blow.

With perfect timing the two policemen rushed through the door, allowing him to turn with a courteous nod.

"Gentlemen, I am sorry," he said. "You are just a trifle too late to witness my experiment. I found it necessary to stop it suddenly and unexpectedly and in an unusual way. . . . However, if you will come around this evening I will have another oscillator attached to this platform and each of you can stand on it. You will, I am sure, find it a most interesting and pleasurable experience. Now you must leave, for I have many things to do. Good day, gentlemen." [16]

When reporters arrived, he blandly told them that he could destroy the Brooklyn Bridge in a matter of minutes if he felt like it.

Years later he told Allan L. Benson of other experiments he had made with an oscillator no larger than an alarm clock. He described attaching the vibrator to a steel link two feet long and two inches thick. "For a long time nothing happened. . . . But at last . . . the great steel link began to tremble, increased its trembling until it dilated and contracted like a beating heart—and finally broke!" [17]

Sledgehammers could not have done it, he told the reporter; crowbars could not have done it, but a fusillade of taps, no one of which would have harmed a baby, did it.

Pleased with this beginning, he put the little vibrator in his coat pocket and went out to hunt a half-built steel building. Finding one in the Wall Street district, ten stories high, with nothing up but the steelwork, he clamped the vibrator to one of the beams.

"In a few minutes," he told the reporter, "I could feel the beam trembling. Gradually the trembling increased in intensity and extended throughout the whole great mass of steel. Finally, the structure began to creak and weave, and the steelworkers came to the ground panic-stricken, believing that there had been an earthquake. Rumors spread that the building was about to fall, and the police reserves were called out. Before anything serious happened, I took off the vibrator, put it in my pocket, and went away. But if I had kept on ten minutes more, I could have laid that building flat in the street. And, with the same vibrator, I could drop Brooklyn Bridge in less than an hour."

Nor was this all. He boasted to Benson that he could split the Earth in the same way—"split it as a boy would split an apple—and forever end the career of man." Earth's vibrations, he went on, have a periodicity of about one hour and forty-nine minutes. "That is to say, if I strike the earth this instant, a wave of contraction goes through it that will come back in one hour and forty-nine minutes in the form of expansion. As a matter of fact, the earth, like everything else, is in a constant state of vibration. It is constantly contracting and expanding.

"Now, suppose that at the precise moment when it begins to contract, I explode a ton of dynamite. That accelerates the contraction and, in one hour and forty-nine minutes, there comes an equally accelerated wave of expansion. When the wave of expansion ebbs, suppose I explode another ton of dynamite, thus further increasing the wave of contraction. And, suppose this performance be repeated, time after time. Is there any doubt as to what would happen? There is no doubt in my mind. The earth would be split in two. For the first time in man's history, he has the knowledge with which he may interfere with cosmic processes!"

When Benson asked how long it might take him to split the Earth, he answered modestly, "Months might be required; perhaps a year or two." But in only a few weeks, he said, he could set the Earth's crust into such a state of vibration that it would rise and fall hundreds of feet, throwing rivers out of their beds, wrecking buildings, and practically destroying civilization. To the relief of ordinary citizens, Tesla later qualified his claim. The *principle* could not fail, he said, but it would be impossible to obtain perfect mechanical resonance of the Earth.

As usual, Tesla's comments to the press smack of exhibitionism. But also, as usual, his research was fundamentally sound. He had begun to establish a new science that he called "telegeodynamics," and it was to have important results. He saw that the same principles of vibration could be used to detect remote objects, such as submarines or ships. By using mechanical vibrations with the known constant of the Earth, he also hoped to learn how to locate ore deposits and oil fields. Modern subsurface exploratory techniques were thus presaged.

Tesla agreed with a theory suggested by O'Neill that a battery of gyroscopes, mounted in a region of severe earthquake hazard, could transmit thrusts into the Earth at equally timed intervals, building up resonance in weak strata and releasing the plate pressure before serious quakes could occur. Today there is renewed interest by seismologists in such techniques.

He described (and later tried to interest Westinghouse in developing) a machine embodying the art of telegeodynamics, with which he claimed to have sent six miles through the Earth mechanical waves "of much smaller amplitude than earthquake waves," that lost little of their power with distance. They were not intended to transmit electrical energy but would enable messages to be carried anywhere in the world and received on a tiny pocket set. Such waves could travel without interference from weather. When pressed by reporters to describe his apparatus, he would say only that it was a cylinder of finest steel—suspended in midair by a type of energy which was old in principle but which had been amplified by a secret principle—combined with a stationary part. Powerful impulses impressed upon the floating cylinder would react on the stationary part and through it, on the Earth.

Nothing was to be developed from this concept. All his life, however, Tesla stuck by his guns as to the awesome potential of mechanical resonance and he went on throwing the fear of God (through science) into impressionable New Yorkers. He could walk over to the Empire State Building, he told reporters, "and reduce it to a tangled mass of wreckage in a very short time." The mechanism would be a tiny oscillator, "an engine so small you could slip it in your pocket." Only 2.5 horsepower would be needed to drive the little vibrator. First, he said, the outer stone coating of the skyscraper would be hurled off. Then the whole vast skeleton of steel, the pride and glory of the Manhattan skyline, would collapse. At this point superman would presumably slip the tiny mechanism into his pocket and casually saunter away, perhaps reciting a line or two from *Faust*. Then his critics would rue the day.

Whatever else Tesla may have been trying to invite by making flamboyant statements such as these—the adulation of his followers, the wrath of other scientists, the consternation of officialdom—he was certainly not courting indifference. But then public indifference was the one thing he could least afford. The more so since fate seemed constantly to be thrusting him into direct competition with that master enchanter of the public imagination, the formidable old Wizard of Menlo Park.

12. ROBOTS

The New Year 1898 found Edison and Tesla in a neck-and-neck race to see who could boggle the minds of lesser mortals with the more outrageous claims. News of their doings had spread all the way to San Francisco, where it was reported that Edison now was "credited with announcing that he can photograph thought. Nikola Tesla tells a New York paper that he has 'harnessed the rays of the sun' and will compel them to operate machinery and give light and heat. This invention is still in the experimental stage, but he declares that there is not a possibility of its failure. He has discovered a method of producing steam from the rays of the sun. The steam runs a steam engine which generates electricity...."[1]

Tesla's solar engine was so simple in design, he said, that if it were fully described others might seize the idea, patent it, and control a blessing "which he intends shall be a free gift to the world." He nevertheless permitted Chauncey McGovern of *Pearson's Magazine* to see his invention, which he claimed employed a single secret factor.

In the center of a large room with a glass roof—his solar municipal powerhouse—reposed a huge cylinder of thick glass on a bed of asbestos and stone. Encircling it would be mirrors covered with asbestos coats to refract the rays of the sun into the glass cylinder.[2] The cylinder would always be kept full of water, which would have been treated by a secret chemical process, and which he said was the only complicated part of the system.

All day long while the sun shone, with the chemical treatment making the water easily subject to heat, steam would be produced to run ordinary steam engines. These in turn would generate electricity for home and factory—enough, indeed, to supply a surplus, to be stored for cloudy days.

The inventor said he fully expected to be ridiculed for having devised a system so simple. The cost of generating such energy would be minimal and he believed—contrary to the experience of subsequent generations—that it should be easy to perfect batteries that could store a whole year's supply of electricity against possible accidents in the generating machinery. The system, he declared, would be a "great deal less artificial than for men to delve down into the bowels of the earth at so much trouble and loss of life in order to get a few handfuls of coal to run an engine a short time and then to make spasmodic return trips for more." Indeed, he hoped to see his solar engine replace not only coal, but wood and every other source of motive power, heat, and light.

Getting his inventions into working form was becoming an ever more serious problem for Tesla, laboring as he did almost alone and besieged with an incessant distracting flight of new ideas. So far as is known, his solar system was never used commercially. And he was having the same trouble with his new vacuum-tube photography lights.

To Robert Johnson he wrote: "I feel confident I have a light which for photography will be better than sunlight, but I have no spare time to bring it to perfection. . . ." He had recently taken a number of photos of the actor Joseph Jefferson to "vindicate" this mysterious new light. (Five years earlier he had taken, with Jefferson as model, the first photographs ever made with phosphorescent light.[3]) Now *The New York Times* reported, "The art of photography will hereafter be independent of sunlight and will be relieved of the inconvenience and discomfort of the flashlight if Nikola Tesla's claims for his latest development of the vacuum tubes are well founded."[4] The *Electrical Review* declared it the oddest and most unlooked-for development of the vacuum tube.[5] Photographs made with the tube were widely printed in newspapers. But thereafter little was heard about it.

Other kinds of practical inventions also intruded on his mind, warring with his preference for basic research. He received an urgent request from George Westinghouse that he provide a "simple and economical device for converting alternating to continuous (DC) currents. . . ." The

Pittsburgh industrialist was interested in converting current for, among other things, running electric trains. Tesla replied at once that he had given a lot of thought to the problem and had "not one but a number of devices to put on your circuit and for all of them there is a great demand."

He was convinced, and so announced, that with properly built railroad tracks, trains running on AC/DC could safely travel up to two hundred miles per hour. As usual his claim gripped the popular imagination even as it griped his fellow inventors. Westinghouse leased one of Tesla's converters. At around this time he also lent the inventor $6,000 to underwrite other inventions in various stages of development. Although Tesla had little money at this point, he at least had no debts.

In May Prince Albert of the Belgians visited the United States and included Tesla's laboratory on his tour. The experience "astonished" him, he said, adding that the inventor was among those Americans who made the strongest impression on him.

Tesla, never one to underestimate the usefulness of royalty, wired George Westinghouse and suggested he invite the Prince to be a guest in his Pittsburgh home. Westinghouse thought it an excellent idea and did so. Afterward Prince Albert visited the Westinghouse power plant at Niagara Falls, attended by his royal entourage.*

Meanwhile, publisher William Randolph Hearst was adroitly steering the nation toward war with Spain; and a strange concurrence of events was shaping that would cause a Teslian moment of glory to be stolen by one of the inventor's closest friends.

Hearst's man in Havana, Frederick Remington, wired his boss: "Everything is quiet. There is no trouble here. There will be no war. I wish to return." To which the great man replied: "Please remain. You furnish the pictures, and I'll furnish the war."[6]

Hearst saw real battles as a solution to the circulation war then rag-

---

* He was not, however, the first Prince Albert to visit the Falls. In 1860, Bertie, Prince of Wales, later to become King Edward VII of England, went to Niagara as a young man and wished to be pushed across the Falls in a wheelbarrow on a tightrope by a French acrobat. He was restrained.

ing between his New York *Journal* and Pulitzer's New York *World.* His
opening journalistic volleys were aimed at Spain for alleged cruelty to
"the gentle Cuban people." When the battleship *Maine* mysteriously
exploded and sank in Havana harbor, he needed nothing more as an
excuse to lash the country into a mood for vengeance. The U.S.
Congress, yielding to the clamor of the press, by a narrow vote declared
war upon Spain.[7]

Americans, fed by the jingoistic press with daily lies and contrived
crises—which included warnings of imminent invasion by the Spanish
navy of cities along the eastern seaboard—responded with righteous
hysteria.

Spain had not the least desire to take on the United States in a fight
she could not possibly win. Nevertheless the American defense machine
was rolled into action; harbors were fortified to repel the imagined
invader and the fighting forces rallied to the flag.

Chauncy Depew, former Secretary of State for New York, gave it as
his opinion that America would never have declared war against Spain
had the matter been left to President McKinley, rather than to a Congress
responsive to the people's mood. And British ambassador James Bryce,
horrified by such irrational preparations and by the lies he read in the
newspapers, said he hoped the country's attitude would not leave a permanent streak of bullying and jingoism in the national character. To this
The New York Times retorted loftily that interceding on behalf of
"oppressed womanhood" could scarcely be interpreted as bullying jingoism. This was a reference to Hearst's romantic crusade to charge to the
rescue of a Cuban rebel known to his American readers only as Miss
Cisneros.

With patriotism pounding in the veins of every loyal son, gestures of a
heroic dimension began to be made even by millionaires. Hearst, for example, sent a letter to the President of the United States: "Sir: I beg to offer to
the United States, as a gift, without any conditions whatsoever, my steam
yacht *Buccaneer.*" In his same "no strings" letter, the publisher requested
that he be given a position in command on his boat. The Navy prudently

accepted the craft but declined the skipper. J. Pierpont Morgan rather more thoughtfully offered to *sell* his yacht, *Corsair*, to the government.

One spring evening in the midst of this national furor, Tesla and the Johnsons, accompanied by their daughter Agnes and handsome naval Lieutenant Richmond Pearson Hobson, dined at the Waldorf-Astoria. It was the Johnson daughter's debut into adult society and a last little fling for Lieutenant Hobson before he bade good-bye to Tesla in his laboratory and vanished on a secret Navy assignment. Almost at once a reporter from the Philadelphia *Press*, as the card in his hatband announced, appeared at the laboratory door.

"I hear you have a wireless device that will communicate with warships one hundred miles away, Dr. Tesla," he said.

"That is true," said the inventor. "But I cannot give you the details. One reason I cannot tell you just what my machine is, is that if it can be used on our ships it will give us an advantage; and I shall be proud to have been of so much use to my country."

"Then you consider yourself a good American?" probed the reporter.

"I, a good American? I was a good American before I ever saw this country. I had studied its government; I had met some of its people, I admired America. I was at heart an American before I thought of coming here to live."

As the reporter scribbled, Tesla expanded.

"What opportunities this country offers a man! Its people are a thousand years ahead of the people of any other nation of the world. They are big, broadminded, generous. I could not have accomplished in any other country what I have here."[8]

He meant it. It was all true. Forgotten were the times when he had been cheated by Edison and his managers and other businessmen, when leading American scientists had derided his polyphase system, when they had laughed at his predictions. That was the way it went sometimes. But it was also true that he was hoping, after an impending exhibition at Madison Square Garden, to interest the government in his very latest wonders.

"The American people are quick to hold out a helping hand and give recognition," he continued. "Yes, I am as good an American as there is. I have nothing to sell the government of the U.S. If it needs my services in any way it is welcome to them."⁹

It was not on the whole, however, a comfortable time for a man of dark complexion and foreign accent to be an American. Hometown "spy-hunts" were just then a popular diversion. Police tended to look the other way if they saw a luckless Spanish-American citizen being beaten up in an alley. Sometimes the "spies" were taken in and grilled for possible deportation.

Andrew Carnegie reflected a popular yearning when he predicted, "Ere long we shall have a solid English-speaking race, capable of preventing much of the evil of the world."

Teddy Roosevelt impetuously resigned as assistant secretary of the Navy and began recruiting Rough Riders from among the membership list of the Knickerbocker Club. Colonel John Jacob Astor mustered an artillery battery. Cowboys and Sioux Indians rallied to the flag. Meanwhile, riots were reported in Spain and starvation in Cuba. In the end, six times as many U.S. troops would die in Cuba of cholera and typhoid as of Spanish bullets.

The day for which Tesla the inventor had been working and waiting arrived in the midst of martial distractions. The first Electrical Exhibition at Madison Square Garden was late in opening, the railroads having been preempted for the movement of soldiers and military supplies and some of the exhibits therefore having failed to arrive on time. Overshadowed by larger events, the show was almost squeezed out of the newspapers. And to cap it all, the weather was rainy. Even so, fifteen thousand persons showed up.

The demonstration of the world's first radio-controlled robot boat by Tesla failed to make the splash it deserved, not only because it was overshadowed by the war, but because he made the mistake of presenting more than the public could absorb at once. The remarkable stage of development to which he had carried wireless, the forerunner of modern radio, would have been quite enough; but to introduce automation simultaneously, as he did, was probably too great a leap. On that day in 1898 when he demonstrated the common ancestor of modern guided weapons and vehicles, of automated industry, and of robotry, he was introducing an idea for which the world would not be ready for many years.

His first two radio-controlled devices were boats, and one was submersible by remote control. On this initial occasion he showed only the submersible. Commander E. J. Quinby (USN Ret.) who, during World War II, was in charge of electronic weapons research for the Navy at Key West, Florida, has written of visiting Tesla's historic exhibit when he was a child: "I was there with my father, quite fascinated, but also quite unaware that I was witnessing the dawn of space navigation to be realized later, in the following century. Tesla was not using Morse code. He was not transmitting messages in any known language. Nevertheless, he was employing his own coded pulses via Hertzian waves to directly control this pioneer unmanned craft. He encoded the visitors' commands, and the vessel's receiver decoded them automatically into actuating operations.[10]

The full potential of the invention was concealed, in part because Tesla hoped the Navy would seriously consider using it in the war.

"One of the features not revealed," science writer Kenneth M. Swezey later disclosed, "was a system to prevent interference by means of coordinated tuning devices responsive only to a combination of several radio waves of completely different frequencies. Another was a loop antenna which could be completely enclosed by the copper hull of the vessel; the antenna would thus be invisible and the vessel could operate completely submerged."[11]

The inventor did not disclose more than his fundamental idea in his basic patent No. 613,809—a means he had learned to use to protect his discoveries.

What his patents included, but the Madison Square Garden viewers did not see, were specifications for a torpedo boat without a crew, including a motor with a storage battery to drive the propeller, smaller motors and batteries to operate the steering gear, and still others to feed electric signal lights and to raise or lower the boat in the water.[12] Six 14-foot torpedoes were to be placed vertically in two rows so that when one was discharged another would fall into place. Tesla had advised the Navy that he thought such a boat could be built for around $50,000.

He claimed that a few such craft "could attack and destroy a whole armada—destroy it utterly in an hour, and the enemy never have a sight of their antagonists or know what power destroyed them."

When word of this got out he received from Mark Twain, then in Austria, a letter in which the humorist wrote: "Have you Austrian and English patents on that destructive terror which you have been inventing? and if so, won't you set a price upon them and concession me to sell them? I know Cabinet ministers of both countries—and of Germany too; likewise, William II.

"I shall be in Europe a year yet.

"Here in the hotel the other night when some interested men were discussing means to persuade the nations to join with the Czar and disarm, I advised them to seek something more . . . than disarmament by perishable paper. . . . Invite the great inventors to contrive something against which fleets and armies would be helpless, and thus make war thenceforth impossible. I did not suspect that you were already attending to that, and getting ready to introduce into the earth permanent peace and disarmament in a practical and mandatory way.

"I know you are a very busy man but will you steal time to drop me a line?"[13]

But the concept was too advanced and those in charge of American defense declared it an impossible dream. Even officials who had observed the midget naval maneuvers in the tank proclaimed it a mere "laboratory experiment" that could never be extended to actual battle conditions.

Tesla's Madison Square Garden demonstration undoubtedly was the most prophetic event at the show, but other inventors also provided displays to bemuse the public. Marconi, without acknowledgment, used a

Tesla oscillator to demonstrate how mines could be blown up by firing a "Cuban dynamite gun" with Marconi's Wireless Telegraphy. And Edison demonstrated what would become his folly, the Magnetic Ore Separator.

Pupin, president of the New York Electrical Society, Edison, and Marconi, a powerful and brainy trio, were now joined by their faith in the financial possibilities of commercial wireless and by three ambitions as great as Tesla's own. One other thing they shared was a growing resentment of Tesla's success.

Tesla and Johnson followed the news of wartime maneuvers and naval encounters from day to day in hope of learning something of the mysterious mission of their friend Hobson. Nothing had been heard directly from him since his abrupt departure in early May.

In the first part of June Spanish Admiral Cervera, whose whereabouts had been the subject of wild speculation in the American press, slipped his vessels into Santiago harbor for coal. An American fleet of superior size moved in. And on the flagship *New York,* unknown to his family and friends at home, was Lieutenant Hobson. He had been thoroughly trained in gunnery and in the handling of explosives.

A desperate scheme, almost a suicide mission, was hatched to bottle up Cervera's fleet. The idea was to sink a ship across the narrowest part of the harbor mouth. The old collier *Merrimac* was chosen and fitted with torpedoes to blow her own hull out. Lieutenant Hobson, at twenty-eight, was chosen to head the mission with a crew of six volunteers.

At 1:30 A.M. on a night of shadowed moonlight the lieutenant and his crew buckled on cork lifebelts over long underdrawers. Armed only with pistols, they moved the old coaling boat slowly toward the harbor mouth.

Hobson reports in a book he later wrote that he said to his gunner's mate, "Charette, my lad, we're going to make it tonight. There is no power on earth that can keep us out of the channel."

At the moment of that ill-timed prediction, a Spanish searchlight picked them out, and the Spanish opened fire. A shell hit their pilot house. Hobson tried to touch off the torpedoes. Only two of them responded, the others having been defectively wired. In short order Spanish fire reduced the *Merrimac* to a sinking wreck—but in a position that failed to block the narrows.

Hobson and his men in their early-day frogmen suits, leaped into the sea and swam to a catamaran that had floated from their deck. But just as they were clambering aboard, a Spanish launch manned by armed soldiers pulled alongside.

Hobson records that as he stared up into their guns he thought, "Despicable cowards! Do they mean to shoot us down in cold blood? If they do, a brave nation will hear of this and call for an account."

It came as something of an anticlimax, therefore, when Admiral Cervera, who was himself on the launch, took the Americans to a Spanish fortress where he treated them with great courtesy and soon exchanged them for Spanish prisoners.

When this feat of derring-do hit American newspapers, they carried little else for days. Hobson was lionized almost to the same degree that Charles Lindbergh would be much later, after flying the Atlantic. Tesla was filled with pride for his friend and delighted when Hobson was sent home for a round of public appearances across the country to rally greater enthusiasm for the war. Tesla and Johnson took the young officer to Delmonico's for a promised celebration and referred to him frequently as "the hero."

Later it greatly amused the inventor to read of how women swarmed over Hobson wherever he went. In Chicago the hero spotted two female cousins he knew and kissed them, which touched off the crowd, causing every woman to demand her due. In Denver he was mobbed again and, according to the press, had to kiss five hundred more. To cap this saccharine frenzy a candy manufacturer announced that he was bringing out a caramel to be called a "Hobson's Kiss."

Tesla was sharply reminded of reality by his bookkeeper, George Scherff, who pointed to the fact that money was running out and that his inventions were not being completed. There were potentially useful items that people needed, he said. For example, doctors and the ailing kept asking for the Tesla Pad—a heat-treating device he had worked on but not perfected for the market.

But where was he to find the time to develop such things?

He enjoyed a rare flurry of socializing with the Johnsons in the winter of 1898 and turned down the usual number of invitations.

On November 3, he wrote to "Dear Kate" saying he was glad she had accepted his invitation for Saturday and adding: "Though a day of plebeians—drummers, grocerymen, Jews,* and other social trilobites, the prospect is nevertheless delightful."[14]

In his invitation he added that a month's income would go on their dinner, but even so, "do not fear it will be extravagant, for just now there is a temporary ebb in my private fortune . . . but soon I am to be a multimillionaire and then good-bye to my friends on Lexington Avenue!"

Shortly afterward, invited to dinner by Katharine and asked to suggest a partner, he predictably named Marguerite. "If she would come," he said, "I know I would."

On December 3, Hobson arrived back in Manhattan, and another celebration was planned. Tesla wrote to Katharine saying, "I am glad. . . . Now we can have that dinner." He suggested that "afterwards we could adjourn to the laboratory," and mentioned a certain lady "who is crazy to see Hobson." Describing her as a great celebrity yet keeping her identity as a surprise, he said he knew how "the Filipovs hunger after such people." And he added, "I do not want to say anything disparaging of a lady, but for my taste she is simply—well, I think you will look more splendid than ever. I warn you she is apt to come in a scarlet décolleté but is a great artist and she must be permitted the latitude. . . . I will sandwich her between Luka and Hobson and wedge you between the hero and myself. . . ."

• • •

* Tesla's anti-Semitism appeared sporadic and was unusual among gentiles of his time. Once he called one of his secretaries to him and hissed as if it were a revealed truth, "Miss! Never trust a Jew!"

Tesla's claims for his first robot vehicles soon came under attack by fellow scientists. Thus "An Inquiry About Tesla's Electrically Controlled Vessel," by N. G. Worth, appeared in the *Electrical Review*, the author expressing his opinion that the method of control could be counter-influenced by the enemy.[15]

Tesla wrote to Johnson at *Century* urging him to make no response on his behalf:

"I know that you are a noble fellow and devoted friend and, noting your indignation at these uncalled-for attacks, I am afraid that you might give it expression. I beg you not to do it under any condition, as you would offend me. Let my 'friends' do their worst, I like it better so. Let them spring on scientific societies worthless schemes, oppose a cause which is deserving, throw sand into the eyes of those who might see—they will reap their reward in time. . . .

"I could easily refute the statements contained therein, merely by referring to expressions of such men as Lord Kelvin, Sir William Crookes, Lord Rayleigh, Roentgen and others, which bear testimony of the high esteem and appreciation of my labors by these men. But I disdain to do so, because the attack was too undignified to deserve notice. . . ."[16]

Under the heading "Science and Sensationalism," the journal *Public Opinion* also criticized his work and methods.[17]

Much later, in his brief autobiography, Tesla disclosed that he had begun active work on building remotely controlled devices in 1893, although the concept had occurred to him earlier. During the next two or three years he had built several mechanisms to be actuated from a distance and showed them to laboratory visitors, but the destruction of the laboratory by fire had interrupted these activities.

"In 1896," he wrote, " . . . I designed a complete machine capable of a multitude of operations, but the consummation of my labors was delayed until 1897. . . . When first shown in the beginning of 1898, it created a sensation such as no other invention of mine has ever produced."

His basic patent was obtained in November, only after the examiner in chief had come to New York and witnessed the performance of his vessel, for he had claimed it seemed unbelievable.

"I remember that when later I called on an official in Washington, with a view of offering the invention to the Government," Tesla wrote, "he burst out in laughter upon my telling him what I had accomplished. Nobody thought then that there was the faintest prospect of perfecting such a device."[18]

These first robots, he wrote in 1919, he had originally considered crude steps in the evolution of the art of teleautomatics. As he had conceived it: "The next logical improvement was its application to automatic mechanisms beyond the limits of vision and at a great distance from the center of control, and I have ever since advocated their employment as instruments of warfare in preference to guns. . . . In an imperfect manner it is practicable, with the existing wireless plants, to launch an aeroplane, have it follow a certain approximate course, and perform some operation at a distance of many hundreds of miles."[19]

He recalled that as a student in college he had conceived of a flying machine quite unlike the present ones.

"The underlying principle was sound but could not be carried into practice," he wrote, "for want of a prime-mover of sufficiently great activity. In recent years I have successfully solved this problem and am now planning aerial machines devoid of sustaining planes, ailerons, propellers, and other external attachments, which will be capable of immense speeds and are very likely to furnish powerful arguments for peace in the near future."[20]

The futuristic aircraft that he conceived of and illustrated was to be guided either mechanically or by wireless energy.

"By installing proper plants it will be practicable to project a missile of this kind into the air and drop it almost on the very spot designated, which may be thousands of miles away. But we are not going to stop at this. Teleautomata will be ultimately produced, capable of acting as if possessed of their own intelligence, and their advent will create a revolution."[21]

As early as 1898 he had also proposed to manufacturers the production of an automated car which, "left to itself, would perform a great variety of operations involving something akin to judgment. But my proposal was deemed chimerical at that time and nothing came from it."

Conceiving of robots as having many uses besides war, he believed their greatest role would lie in peaceful service to humanity. He later described his 1890s activity to Professor B. F. Meissner of Purdue University: "I treated the whole field broadly, not limiting myself to mechanisms controlled from distance but to machines possessed of their own intelligence. Since that time I had advanced greatly in the evolution of the invention and think that the time is not distant when I shall show an automaton which, left to itself, will act as though possessed of reason and without any wilful control from the outside. Whatever be the practical possibilities of such an achievement, it will mark the beginning of a new epoch in mechanics."

He added: "I would call your attention to the fact that while my specification, above mentioned, shows the automatic mechanism as controlled through a simple tuned circuit, I have used individualized control; that is, one based on the co-operation of several circuits of different periods of vibration, a principle which I had already developed at that time and which was subsequently described in my patents 723,188 and 723,189* of March, 1903. The machine was in this form when I made demonstrations with it in 1898 before the Chief Examiner (of Patents) Seeley, prior to the grant of my basic patent on Method of and Apparatus for Controlling Mechanisms at a Distance."[22]

It was this that Swezey alluded to in his comments on "coordinated tuning devices responsive only to a combination of several radio waves of completely different frequencies."

Inventors of modern computer technology in the last half of the twentieth century repeatedly have been surprised, when seeking patents, to encounter Tesla's basic ones, already on file. Leland Anderson, for example, states that Tesla's priority was first pointed out to him years ago by a patent attorney for a major computer firm with which he was asso-

---

* Tesla's letter to Meissner erroneously listed patent 723,189, the correct number being 725,605, which was issued April 14, 1903.

ciated in a research and development capacity. Anderson writes, "I am puzzled by the reluctance of some in the computer technology field to acknowledge Tesla's priority in this regard in contrast to the adulation given to Messrs. Brattain, Bardeen, and Shockley for the invention of the transistor which made electronic computers a practical reality."[23]

Their patents and the Tesla patents were both directed at applications in the communications field, he notes. Both patents are combined to produce the physical embodiment of a solid-state AND gate. Computer systems contain thousands of logic decision elements called ANDs and ORs. All operations performed by a computer are achieved through a system design utilizing these logic elements.

"Tesla's 1903 patents 723,188 and 725,605," says Anderson, "contain the basic principles of the logical AND circuit element. The simultaneous occurrence of two or more prescribed signals at the input to the device element produced an output from the device element."

Although Tesla's patents used AC signals and today's computers use pulsed DC, the basic principle of a prescribed combination of signals producing an output by virtue of their conjoint action is described.

"Thus," declares Anderson, "the subject early Tesla patents, which were designed to achieve interference protection from outside influences in the command of radio-controlled weapons, have proved to be an obstacle for anyone attempting to obtain a basic logical AND circuit element patent in this era of modern computer technology."[24]

The Nobel Prize was awarded to John Bardeen, Walter H. Brattain, and William B. Shockley in 1956 for their work on developing the transistor, which replaced electronic tubes in many applications. Yet Tesla has only recently been so much as recognized for having pioneered the field.

One of the earliest acknowledgments of the debt owed Tesla in the new technology of remotely piloted vehicles (now universally known in the military as RPVs) appeared in a 1944 *Times* editorial:

"The general principle of controlling apparatus by radio goes back to the early days of what was once called 'wireless.' At the first electrical exposition held in this city over forty years ago Nikola Tesla maneuvered and blew up a model submarine in a tank by radio. There soon followed a score of German, American, English, and French inventors who showed how engine-driven vehicles, torpedoes and ships could be steered by radio waves with never a man on board. . . ."[25]

Yet Tesla, having done so much to introduce the era of automation, felt that he had no time just then to pursue a line of development for which the world was still manifestly unready. His sights were fixed on bigger game—if that were possible. His laboratory in New York was no longer a safe place for his experiments; or, rather, his experiments had become too dangerous for a crowded city.

To Leonard Curtis, a patent attorney who had loyally protected his and Westinghouse's rights during the War of the Currents, he wrote: "My coils are producing 4,000,000 volts—sparks jumping from walls to ceilings are a fire hazard. This is a secret test. I must have electrical power, water and my own laboratory. I will need a good carpenter who will follow instructions. I am being financed for this by Astor, and also Crawford and Simpson. My work will be done late at night when the power load will be least."[26]

Curtis, who was associated with the Colorado Springs Electric Company, immediately set to work on the inventor's problem. His solution would have far-reaching consequences.

13. HURLER OF LIGHTNING

Leonard Curtis's reply from Colorado Springs could not have brought better news: "All things arranged, land will be free. You will live at the Alta Vista Hotel. I have interest in the City Power Plant so electricity is free to you."

Tesla, overjoyed, threw himself into detailed preparations, especially the ordering of machinery that would have to be shipped. Meanwhile, Scherff and his shop assistant, Kolman Czito, were called upon to labor almost around the clock for a major move of laboratory equipment.

Of paramount importance was the reorganizing of his finances. The $40,000 paid to him by Adams for stock in the Nikola Tesla Company had long since been spent. Ten thousand dollars given to him by John Hays Hammond, Sr., the famous mining engineer, had gone to underwrite his wireless and robot work for the Electrical Exhibition. But the drygoods firm of Simpson and Crawford lent another $10,000 to him for ongoing research, and Col. John Jacob Astor, owner of the Waldorf-Astoria Hotel, contributed $30,000 toward the building of the new research station in Colorado Springs.[1]

Once established in Colorado, Tesla intended to devote all his energies toward an immediate dual goal: to develop a worldwide wireless system well ahead of the ambitious Marconi, and to learn how to send energy abundantly and cheaply without wires to the ends of the Earth. No body of knowledge, except that which he had already developed, existed to guide him.

Yet there remained a little time to socialize with his friends, a little time to rekindle jealousy in the adoring Katharine. Marguerite was the pawn in this game, if it was a game.

"Agnes will come by all means," he wrote to Kate, as if she were his social secretary. "And—wouldn't you invite Miss Merington? She is such a wonderfully clever woman. . . . Really I would like to have her with us. . . ."[2]

On the twenty-fifth of March he begged off a date with Luka, "having already accepted an important engagement with an English millionaire." But he described his joy at having at last moved into the fashionable Waldorf-Astoria Hotel after ten years at the "abominable place" that appeared to pride itself more on the quality of being fireproof than on the honor of having a distinguished inventor in residence.[3]

Colonel Astor, at least, felt honored to have him as a guest. And Tesla was instantly at home in his smart new surroundings, where all the important men of Wall Street gathered in the afternoon.

In the flurry before departure he found time to initiate an effort to "get permission from the French government for transmitting energy and establishing communication with France without wires, in view of the coming Exposition. . . ." His reason for this was to be disclosed on his arrival in Colorado.

Tesla departed New York on May 11, 1899, traveling by train and making a stopover in Chicago to demonstrate again his radio-controlled boat. George Scherff was left behind to run the New York laboratory, with precise and lengthy instructions for more equipment to be built, bought, and shipped. Of course Tesla left him with neither adequate money nor a power of attorney to cover the day-to-day expenses. As the inventor saw the matter, when he considered it at all, his staff would soon share in his own wealth and fame.

Arriving at Colorado Springs on May 18, he was taken directly to the Alta Vista Hotel. After examining the creaky elevator, he chose room No. 207 (divisible by three and only one flight up), and left instructions for the maid to deliver eighteen clean towels daily. He said he preferred to do his own dusting.

The land made available to him was about a mile east of Colorado Springs, in the shadow of Pike's Peak. Its main use was grazing pasture for the town's dairy herd. His closest neighbor was to be the Colorado School for the Deaf and Blind, a choice reflecting some discretion. The elevation was 6,000 feet above sea level; the air clear, dry and crackling with static electricity.

To reporters who interviewed him on his arrival, he disclosed that he planned to send a wireless message from Pike's Peak to Paris in time for the Paris Exposition of 1900. The journalists asked whether he meant to send messages from peak to peak. He replied haughtily that he had not come to Colorado to engage in stunts.

He had filed in the preceding decade a whole series of patents related to the wireless transmission of power and messages, beginning with the most basic equipment for the production of high frequencies and high voltages.* He had already built a coil that produced 4 million volts, and now he wanted to go much higher in order to power a device capable of making transmissions on a global scale. The tests were to be made in great secrecy—or, at any rate, as much secrecy as was possible in a small community titillated by the arrival of a famous inventor with mountains of mysterious equipment.

Tesla was directed to a local carpenter named Joseph Dozier, to whom he outlined plans for the experimental station, and the construction began immediately. He then sent the first of an almost continuous stream of wires and letters to Scherff in New York asking that Fritz Lowenstein, his young engineering assistant, be sent west: "He must be here to oversee construction and locate equipment."

During the building of the experiment station the inventor commuted to and from the site each day by buckboard, his long legs sprawled over the sides—not so much from lack of space as in readiness to abandon ship. Tesla trusted horses no more than he did electric elevators. (In time, the horses of Colorado Springs would have equal reason

*In a heterogeneous basic group: No. 454,622 (first patent of the coil named after Tesla), 462,418, 464,667, 512,340, 514,167, 514,168, 567,818, 568,176 568,178, 568,179, 568,180, 577,670, 583,953, 593,138, 609,245, 609,246, 609,247, 609,248, 609,249, 609,251, 611,719, and 613,735, plus the two relating to wireless transmission of power and messages, 645,576 and 649,621, all filed prior to his Colorado experiments. His work in Colorado provided the foundation for several important patents: No. 685,953, 685,954, 685,955 and 685,956 referring to receivers; and for most of them he applied while in Colorado. Shortly after returning to New York he filed another group (see footnote p. 208).

not to trust Tesla, for when he got his powerful magnifying transmitter operating, it would electrify the Earth in all directions, making runaways of the gentlest nags.)

A fence surrounded the weird structure that began to rise from the prairie floor, and this barrier bristled with warnings: "KEEP OUT— GREAT DANGER." When the station was completed an even more ominous quotation from Dante's *Inferno* was posted at the door: "Abandon hope all ye who enter here." It did not take long for the word to spread that the apparatus being built by Mr. Tesla was capable of killing a hundred persons in a single flash of lightning.

The experiment station, which had started out looking like a large square barn, ended up resembling a ship with a towering mast. Extruding from an open section of the roof was a tower that reached eighty feet above the ground. From this metal mast soared another 122 feet into the air. Poised upon its tip was a copper ball three feet in diameter.

Machinery was moved in and assembled as quickly as it arrived on the construction scene. Coils or high-frequency transformers in many shapes and sizes were built. From New York came the specially built two-turn primary circuit that he had had in his laboratory on Houston Street. With its associated circuit interrupters, it would drive his magnifying transmitter.

This transmitter, which he developed in Colorado, he would later claim as his greatest invention. Indeed, it is the Tesla invention that continues to fascinate many of his modern followers the most. Whenever and wherever in recent years phenomena have been detected, resulting from powerful radio signals pulsed at very low frequencies, journalists speak knowingly of the Tesla effect. The Russians, it has been claimed, are using a giant Tesla magnifying transmitter to modify the world's weather, creating extremes of ice and drought. It is said to cause periodic disruption of radio communications in Canada and the United States with attendant brain-wave interference and vague symptoms of physical distress, not to mention sonic booms and almost anything else not otherwise explicable. Indeed, it was this same fabulous invention that Robert

Golka in recent years tried to replicate, with considerable success, at Wendover, Utah, for the study of ball lightning, in conjunction with research in nuclear fusion.

But what exactly was it? Tesla was asked to describe it for *The Electrical Experimenter* in a way that young readers could understand. His explanation (which must have taxed his readers) is tantalizingly vague. "Well, then, in the first place," he wrote, "it is a *resonant transformer* with a secondary in which the parts, charged to a high potential, are of considerable area and arranged in space along ideal enveloping surfaces of very large radii of curvature, and at proper distances from one another thereby *insuring a small electric surface density everywhere so that no leak can occur even if the conductor is bare.* It is suitable for any frequency, from a few to many thousands of cycles per second, and can be used in the production of currents of tremendous volume and moderate pressure, or of smaller amperage and immense electro-motive force. The maximum electric *tension is merely dependent on the curvature of the surfaces* on which the charged elements are situated and the area of the latter."[4]

One hundred million volts, he declared, were perfectly practicable. Such a circuit could be excited with impulses of any kind, even of low frequency, and would yield sinusoidal and continuous oscillations like those of an alternator.

"Taken in the narrowest significance of the term, however," Tesla wrote, "it is a resonant transformer which, besides possessing these qualities, is accurately proportioned to fit the globe and its electrical constants and properties, by virtue of which design it becomes highly efficient and effective in the wireless transmission of energy. Distance is then absolutely eliminated, there being *no diminution in the intensity of the transmitted impulses.* It is even possible to make the actions *increase with the distance from the plant* according to an exact mathematical law."[5]

Once this powerful equipment was built and the inventor began testing he was able to emulate the electrical fireworks of even the wildest mountain storms. When the transmitter was operating, lightning arresters in a twelve-mile radius from his station were bridged with continuous fiery arcs, stronger and more persistent than those produced by natural lightning.

For the first time he kept a careful daily diary in which he recorded every aspect of his research. And because visual effects were useful as well as thrilling, he devoted many hours to photographic experiments.

The equipment Tesla was perfecting would, he hoped, one day be adaptable for commercial use. But first, thousands of observations and delicate adjustments had to be made. He no longer trusted his legendary memory to store such a volume of information. His daily notes referred constantly to experiments that had failed to turn out as expected, and he would ask himself why. This process was at sharp variance with the one he claimed to have used throughout his earlier life. Now middle-aged, he may have felt his memory waning slightly. Certainly he felt driven by the pressures of his self-imposed deadline.

In his Colorado journal his lifelong fascination with visual phenomena is underscored. The flashing lights that he had always experienced on the screen of his mind were dramatically externalized, and his descriptions, among the mass of mathematical formulas, are detailed, loving, almost erotic in their lingering portrayal of the colors and grandeur of his Colorado electrical storms.[6]

Nights when experiments were being made with the magnifying transmitter the prairie sky exploded with sound and color. Even the earth seemed alive and the crash of thunder from the spark gap could be heard for miles. Butterflies were sucked into the vortex of the transmitter coil, which was fifty-two feet in diameter. Awed spectators at some distance from the station told of seeing tiny sparks flying between grains of sand and between their heels and the ground when they walked. They said that at three hundred feet away, arcs an inch long could be drawn from grounded metal objects.[7] Horses grazing or trotting peacefully half a mile away would suddenly go berserk, feeling shocks through their metal shoes.

The inventor and his assistants, working nightly amidst thunder and lightning, stuffed cotton in their ears and wore thick cork or rubber soles on their shoes. Even so, Tesla described a frequent bursting sensation in the ears, something almost as positive as touch, and feared damage to their eardrums. Often the pain and buzzing they felt continued for hours after a test.

Hertz's research of 1888, which confirmed Maxwell's dynamic theory of the electromagnetic field, had convinced scientists that electromagnetic waves propagated in straight lines, like light waves. Therefore it was generally believed that radio transmission would be limited by the curvature of the Earth. Tesla, as we know, believed not only that the globe was a good conductor but that the "upper strata of the air are conducting" and "that air strata at very moderate altitudes, which are easily accessible, offer, to all experimental evidence, a perfect conducting path."

Until recent years this theory of propagation of radio waves was ignored. In the 1950s, however, a number of scientists working on the propagation of very low (3 to 30 kHz) and extremely low (1 to 3000 Hz) electromagnetic waves confirmed Tesla's principle insofar as they apply to low-frequency transmissions. As the world authority on electromagnetic wave theory, Dr. James R. Wait, has observed, Tesla's experiments at Colorado Springs, "predate all other electromagnetic research in Colorado . . . [and] his early experiments have an intriguing similarity with later developments in ELF (extra-low frequency) communications."[8] In fact, the Tesla magnifying transmitter was the first in the world powerful enough to create ELF resonance in the earth-ionosphere wave guide.

He was equally prescient in a prediction made at this time that the Earth resonates at 6, 18, and 30 Hz. He later tried to verify this with equipment he built on Long Island, but not until the 1960s would the experiments that he had wanted to carry out be made by others. It was then found that Tesla had been remarkably close to the mark: The Earth resonates at 8, 14, and 20 Hz.

Since his wireless power-transmission concept involved Earth resonance, the closer he could bring his operational frequency to that of the Earth the better it would be for producing very large movements of power in his system. But low frequencies presented a difficult problem insofar as the length of his secondary winding was concerned. For example, for his magnifying transmitter, which operated at 50 kHz, the winding length was approximately 0.9 miles. At 500 Hz, the length would have had to be ninety miles.

Progress reports and requests for shipments were scorching the telegraph wires between Tesla and Scherff. Regular freight was too slow for the inventor, so he ordered Scherff to use expensive railway express. The presence of Kolman Czito was commanded. Tesla wrote Scherff, advising him that Czito's salary of $15 per week was to be paid to his wife. Soon he was able to report, "Czito has just arrived and I was glad to see a familiar face again. He looks a little too fat for the work I expect of him."

There was also discussion by telegraph about the two hundred bottles Tesla had ordered and about the eight-foot balloons that, according to Scherff, Mr. Myers feared "would not rise at the altitude where you are if it should be windy." [9] The balloons were to hoist stationary antennae into the high thin air. Eventually they were designed by a professional at $50 each and were to be filled only two-thirds full (probably with hydrogen) to avoid breakage at a great height.

Scherff, knowing of his hunger for news, kept him advised of every detail of the progress at home and especially informed him of the movements of Colonel Astor, his major financial backer. He also reported on the activities of Marconi and on matters related to Tesla's European patents.

Busy as both men were, they found time to exchange the slightest tidbits of gossip or instruction. "Mr. L." said Scherff, "has been coming to the shop intoxicated and making many errors in his drilling." And Tesla admonished, "Tell Mr. Uhlman not to write *yours truly,* but *sincerely;*" and signed his own letter to Scherff, "Yours sincerely." He added an anxious P.S.: "Has my friend JJA [Astor] called?"

To Scherff he enlarged on the problems of security and promised him reflected glory: "Do everything you can intelligently keeping the interest of my efforts in view and be particularly careful to any press representatives. I do not want you to say anything except what I state here. I think when I come back I shall have something to say.... You must all be as part of myself, then I shall pull you with me to success."[10]

On August 16 he wrote to "My dear Luka" to thank him for his poem, "Dewey at Manila," which was "simply great," and added: "I wish you could see the snowdrops and icebergs of Colorado Springs! I mean those that float in the air. They are sublime, next to your poems, Luka, the finest things on Earth! Kind regards to all from your Nikola."

But a little later he again wrote Johnson on a less ecstatic note. "The wireless torpedo got on the scene just a trifle too late and Dewey slipped into the gallery of immortal conqueror—but it was a close shave! Luka, I see every day that we are both too far ahead of our time! My system of wireless telegraph is buried in the transactions of a scientific society, and your great poem on the heroes of Manila did not even as much as save Montojo, and just as my enemies maintain that I am merely writing ideas of others, so yours will say that it is because of your poem that Montojo was condemned!

"But we shall continue in our noble efforts, my friend, not minding the bad and foolish world, and sometime . . . I shall be explaining the principles of my intelligent machine (which will have done away with guns and battleships) to Archimedes, and you will read your great poems to Homer...."[11]

Scherff wrote: "The New York *Herald* continues to boom Marconi...."

For all his worries about the project, Tesla was finding Colorado's weather and atmosphere exhilarating. His vision and hearing, both of which were always acute, responded to an extraordinary degree to the clarity of the air. The climate was ideal for his observations. The sun's rays were fiercely intense, the air dry, and the frequent lightning storms of almost inconceivable violence.[12]

In mid-June, with all of his equipment installed and preparations for various tests going forward, he arranged one of his receiving transformers with a view to determining experimentally the electrical potential of the globe. In accordance with a careful plan, he wished to study its periodic and casual fluctuations.

He placed a highly sensitive device controlling a recording instrument in his secondary circuit, and, with the primary connected to the ground, he placed the secondary on an elevated terminal. This produced a surprising result: The variations of electrical potential gave rise to electrical surgings in the primary; these generated secondary currents, which in turn affected the sensitive recorder in proportion to their intensity.

"The earth," Tesla later reported in an article, "was found to be, literally, alive with electrical vibrations, and soon I was deeply absorbed in this interesting investigation. No better opportunity for such observations as I intended to make could be found anywhere."[13]

The natural lightning discharges in this part of Colorado were very frequent and sometimes of great violence, on one occasion about twelve thousand discharges occurring within two hours, all within thirty miles of Tesla's laboratory. Many of them he described as resembling gigantic trees of fire with their trunks upside down. And toward the end of June he noticed a curious phenomenon: His instruments were being affected more strongly by discharges occurring at a great distance than by those nearby. "This puzzled me very much," he wrote. "What was the cause?"

One night while he was walking home across the prairie with the stars glowing coldly above, a possible explanation came to him. The same idea had occurred to him years before when he was preparing his lectures for the Franklin Institute and the National Electric Light Association, but then he had dismissed it as absurd and impossible. "I banished it again," he wrote. "Nevertheless, my instinct was aroused and somehow I felt that I was nearing a great revelation."[14]

14. BLACKOUT AT
COLORADO SPRINGS

It was on the 3rd of July [1899]—the date I shall never forget—when I obtained the first decisive experimental evidence of a truth of overwhelming importance for the advancement of humanity."[1]

At dusk of that day Tesla had watched a dense mass of strongly charged clouds gathering in the west. Soon the usual violent storm broke loose "which, after spending much of its fury in the mountains, was driven away at great speed over the plains."

He noticed heavy and persistent lightning arcs forming almost at regular time intervals. Prepared with a recording instrument, he noted that its indications of electrical activity became fainter and fainter with the increasing distance of the storm, until they finally ceased altogether.

"I was watching in eager expectation," he noted in the diary. "Surely enough, in a little while the indications again began, grew stronger and stronger, and, after passing through a maximum, gradually decreased and ceased once more. Many times, in regularly recurring intervals, the same actions were repeated until the storm which, as evident from simple computations, was moving with nearly constant speed, had retreated to a distance of about 300 kilometers. Nor did these strange actions stop then, but continued to manifest themselves with undiminished force."[2]

Soon Tesla felt sure of the true nature of the "wonderful phenomenon. No doubt whatever remained: I was observing stationary waves."[3]

He summed up the implications of this discovery thus: "Impossible as it seemed, this planet, despite its vast extent, behaved like a conductor of limited dimensions. The tremendous significance of this fact in the transmission of energy by my system had already become quite clear to me.

"Not only was it practicable to send telegraphic messages to any distance without wires, as I recognized long ago, but also to impress upon the entire globe the faint modulations of the human voice, far more still, to transmit power, in unlimited amounts to any terrestrial distance and almost without loss."

Tesla visualized the Earth as an extremely large container holding an electrical fluid which resonance caused to be formed into a series of waves frozen in position. It was now certain, he wrote, that stationary waves could be produced in the Earth with an oscillator. *"This is of immense importance."*[4] He already knew that power transmission and the sending of intelligible messages to any point of the globe could be achieved in two radically different ways: either by a high ratio of transformation or by resonant rise. From the tests with electrical oscillators he now concluded—and so noted in his diary—that power transmission would be best served by the first method, but that where a small amount of energy was needed, as with radio, "the latter method is unquestionably the better and simpler of the two."[5]

Later, leading scientists would erroneously criticize him for having made no distinction between the two functions. And, in keeping with his policy of secrecy, he did not trouble to enlighten them. But before Tesla would apply his theories practically, he first had to perfect his equipment. The test for which he next prepared called for millions of volts and tremendously heavy currents. No past experience could prepare him for what might happen, except in a general way. Bolts of his man-made lightning were bound to explode from the top of the two-hundred-foot mast and tower structure, but whether they would kill the experimenters and burn down the station was a risk they would have to take.

On the appointed night he dressed neatly and carefully in his black Prince Albert coat, donned gloves and a black derby hat, and arrived at the station to find courageous Czito already waiting. The latter would man the switch, giving Tesla the opportunity to observe effects from the doorway of the laboratory. It was important for him to watch both the giant coil in the center of the room and the copper ball on the mast.

When all was ready, he shouted, "Now!"

It had been prearranged that on the first test the switch was to be closed for only a single second. Accordingly, Czito slammed it in, watched the second hand on his pocket watch, and almost instantly pulled it out. The effects in that brief instant were rewarding: threads of fire had crowned the secondary coil and electricity snapped above.

For the main event Tesla wanted to watch from outside where he would have a clear view of the mast and ball.

"When I give you the signal," he told Czito, "I want you to close the switch and leave it closed until I give you the signal to open it."

In a moment he called, "Now! Close the switch!"

Czito followed orders and stood poised to pull it out again on command. The vibration of heavy current surging through the primary coil made the ground feel alive. There came a snap and a roar of lightning exploding above the station. A strange blue light filled the interior of the barnlike structure.

Czito looked up to see the coils a mass of surging, writhing snakes of flame. Electrical sparks filled the air and the sharp smell of ozone stung his nostrils. Lightning exploded again and again, building to a crescendo, and still Czito waited for the order to yank open the switch. Unable to see Tesla from his post, he began to wonder if the inventor had been struck by lightning and lay injured or dead outdoors. To continue seemed madness. In another moment he feared the walls and roof of the station would be aflame.

Tesla, however, was neither injured nor dead. He was frozen in a paroxysm of bliss. From where he stood he could see the lightning bolts shooting 135 feet from the top of the mast, and as he later learned, the thunder was being heard fifteen miles away in Cripple Creek. Again and again the lightning surged and crashed. Sublime! Had ever a human being felt more in tune with the gods? How long he stood there he had no idea. Later it turned out to have been only about one minute.

But suddenly, inexplicably, all was silent. What could have happened? He shouted to Czito: "Why did you do that? I did not tell you to open the switch. Close it again quickly!"

Czito, however, had not touched the switch. The power was dead. God in His mercy had sent him a reprieve.

Tesla rushed to a telephone and called the Colorado Springs Electric Company. He began remonstrating and pleading. They had cut off his power, he charged, and must restore it at once.

The reply from the powerhouse was curt and to the point.

"You've knocked our generator off the line, and she's now on fire!"[6]

Tesla had overloaded the dynamo. The town of Colorado Springs was in darkness. As soon as the fire was extinguished a standby generator was put into service, but Tesla's request to be served by it was brusquely denied.

Determined to continue his experiments, he offered to take a team of skilled workmen to the powerhouse and repair the main generator at his own expense. The offer was accepted. Within a week the repairs had been made, and Tesla was once more provided with electricity.

Thereafter his experiments progressed smoothly. Scherff continued to ship new apparatus to him through the icy Colorado fall and winter. To encourage the inventor he wrote, "Mr. Lowenstein has told Mr. Uhlman and me something of your wonderful work, and we know that, instead of a century, you are a thousand years ahead of others."

Unfortunately, we have only an imperfect idea of some of the things Tesla attempted—and, for all we know, accomplished—during this period. His diary notes and later writings are often maddeningly unenlightening. For example he appears at one point to have been experimenting with the production of some kind of potent ray. Among the items rushed to him by express were four double-focus Roentgen tubes with thick platinum targets; and one journal entry reads: "Arrangements with single terminal tube for production of powerful rays. There being practically no limit to the power of an oscillator, it is now the problem to work out a tube so that it can stand any desired pressure. . . ."[7] The exact purpose or results of these experiments are unknown, but for further information, please see chapters 29 and 30.

The general thrust of his inquiries is, of course, clear. He tested high-power oscillators, wireless transmission of energy, the reception and transmission of messages, and the related effects of high-frequency electric fields.

Whatever their nature, his experiments seldom lacked glamor. Despite the warning signs he had put up on the fences and building, he had been disturbed by neighborhood boys peering in through a single window in the rear. Tesla had it nailed up. As a result of this, he came as near to being killed as ever before in his risk-taking life.

"It was a square building, in which there was a coil 52 feet in diameter, about nine feet high," he later recalled. "When it was adjusted to resonance, the streamers [of electricity] passed from top to bottom and it was a most beautiful sight. You see, that was about fifteen hundred, perhaps two thousand square feet of streamer surface. To save money I had calculated the dimensions as closely as possible and the streamers came within six or seven inches from the sides of the building."[8]

The main switch for handling the heavy currents had proved hard to pull. To make its operation easier, Tesla had installed a spring that would cause it to snap closed at the merest touch. This innovation was soon revealed as more convenient than safe.

On the day in question Tesla had sent Czito downtown and was experimenting alone. "I threw up the switch and went behind the coil to examine something. While I was there, the switch snapped in, when suddenly the whole room was filled with streamers, and I had no way of getting out. I tried to break through the window but in vain as I had no tools, and there was nothing else to do than to throw myself on my stomach and pass under.

"The primary carried 50,000 volts, and I had to crawl through a narrow place with the streamers going. The nitrous acid was so strong I could hardly breathe. These streamers rapidly oxidize nitrogen because of their enormous surface, which makes up for what they lack in intensity. When I came to the narrow space they closed on my back. I got away and barely managed to open the switch when the building began to burn. I grabbed a fire extinguisher and succeeded in smothering the fire...."[9]

He wrote a Dear Luka letter in which he metaphorically alluded to having tamed a wildcat and to being a mass of bleeding scratches.

"But in the scratches, Luka," he wrote, "there lives a mind—a MIND! Well, I do not want to say much, but ...

"I have made splendid progress in a number of lines but—how grieved I was to find that a number of my confreres of wireless telegraphy—of the syndicating kind—have been indulging in an awful lot of lying! Not a single of the contentions they have brought forth is true and my system, Luka, is used—pure and simple—without the slightest departure. ..."[10]

This referred to Marconi who, working with the English Post Office electrician, William Preece, had sent a wireless signal eight miles across the Bristol Channel two years earlier and who now, in 1899, had just repeated this performance across the English Channel.

Edison, reminded of his own unpromising experiments sixteen years before, now would begin to wonder if *he* might have the basis for a lawsuit against the young Italian. As it happened, he later would be awarded $60,000 from the Marconi Wireless Telegraph Company for his patent. But in truth, perhaps because of his deafness, Edison had never really believed that the "radio craze" would last.

As for Tesla in Colorado, he confided to Robert Johnson that he was absolutely sure he would transmit a message to the Paris Exposition of 1900 without wire—"my greeting to the crazy French!" He closed on a familiar note: "I have not yet found time to carry out my promise of becoming a millionaire, but I shall do so at the earliest opportunity. ..."[11]

What exactly did Tesla achieve during his sojourn in Colorado Springs? Certainly all the mystery, the furious activity, the considerable expense, and the periodic theatrical effects failed to produce any single practical invention—if by practical one means a telephone or a better bobbin. Judged by "Edisonian" standards, however, one might as well protest that Einstein invented no electric dishwasher.

But did Tesla during this period, then, make significant contributions to new knowledge? The answer is *yes*. Scholars do not know and may never know the full range of his explorations, and there is the further problem that he often did not follow up on his intuitions, theories, and preliminary experiments to the point of verification. But he certainly made significant fundamental contributions as his scientific successors in several fields continue to discover. (See chapter 30.)

The eminent Yugoslav physicist Dr. Aleksandar Marinčić points out that today when we have proof of the Earth's resonant modes and know that certain waves can propagate with so little attenuation that standing waves can be set up in the earth-ionosphere system, "we can judge how right Tesla was when he said that the mechanism of electromagnetic wave propagation in 'his system' was not the same as Hertz's system with collimated radiation." In his introduction to Tesla's *Colorado Springs Notes,* Dr. Marinčić observes that the scientist could not, however, have known "that the phenomena he was talking about would only become pronounced at very low frequencies"; and he surmises that further study of Tesla's writings "will reveal some interesting details of his ideas in this field." The diary especially throws light on his part in the development of radio, and there is no longer a question of his mastery of wireless transmission as early as 1893.

In part, however, scholars can only try to reconstruct what Tesla *thought* he had accomplished.

With his giant oscillator he believed he had set the Earth in electrical resonance, pumping a stream of electrons (at that time, a flow of electricity) into it at a rate of 150,000 oscillations per second. The resulting pulsations had a wavelength of about 6,600 feet. Tesla concluded that they expanded outward over the bulge of the Earth, first in increasing circles and then in ever smaller ones yet with growing intensity, and converged at a point on the globe directly opposite from Colorado Springs—that is, slightly west of the French islands of Amsterdam and St. Paul in the Indian Ocean.

Here, according to his experimental results, a great electrical "south pole" was created with a stationary wave that rose and fell in unison with his transmissions from his "north pole" at Colorado Springs. Each time the wave receded, it was reenforced and sent back more powerfully than before to the antipode.[12]

Had the Earth been capable of perfect resonance, the results could have been catastrophic, but since it was not, the effect, he believed, was merely to make available at any point on the Earth energy that could be drawn off with a simple piece of equipment. This would include the elements of a radio tuning unit, a ground connection, and a metal rod the height of a house. Nothing more would be needed to absorb household electricity from the waves rushing back and forth between the electrical north and south poles. He did not, however, satisfactorily prove this claim, let alone apply it. Nor has anyone else.

With his magnifying transmitter he had produced effects at least in some respects greater than those of lightning. The highest potential he reached was about 12 million volts, which is insignificant compared to that of lightning, yet far higher than anyone else produced for many decades thereafter. What he considered *more* significant, however, was that he obtained in his antenna current strengths of 1100 amperes. The biggest wireless plants for many years thereafter used only 250 amperes.*

One day, working with such currents, he succeeded, to his surprise, in precipitating a dense fog. There was a mist outside, but when he turned on the current the cloud in the laboratory became so dense that he could not see his hand inches in front of his face. From this he concluded that he had made an important discovery. "I am positive in my conviction," he said later, "that we can erect a plant of proper design in an arid region, work it according to certain observations and rules, and by its means draw from the ocean unlimited amounts of water for irrigation and power purposes. If I do not live to carry it out, somebody else will, but I feel sure that I am right."

This idea too went into his legacy of unfinished business, and to this date no one has implemented it.

It has been reported by various writers that during his power trans-

---

* When accepting the Edison Medal in 1917, Tesla recollected he had reached a potential of 20 million volts.

mission experiments in Colorado, Tesla succeeded in lighting up a bank of two hundred 50-watt incandescent lamps wirelessly, at a distance of twenty-six miles from his station. In his own writings, however, no such claim was ever made, nor is there other evidence that he did so. What he actually wrote was that, by use of the magnifying transmitter, he had passed a current around the globe sufficient to light more than two hundred incandescent lamps.

"While I have not as yet actually effected a transmission of a considerable amount of energy, such as would be of industrial importance, to a great distance by this new method," he would write on returning East, "I have operated several model plants under exactly the same conditions which will exist in a large plant of this kind, and the practicability of the system is thoroughly demonstrated."[13] He also wrote that he had observed the transmission of signals up to a distance of 600 miles.

That was as specific as he cared to be on the subject. Two other remarkable scientific achievements, however, were to result from his months of concentrated research in Colorado.

In a diary entry dated January 3, 1900, after describing the taking of some laboratory photographs, he mentioned watching the formation of sparks into streamers and "fireballs."[14] Ball lightning, or fireballs, is a phenomenon that has fascinated and baffled scientists from ancient times to the present. Fireballs are mentioned on Etrurian monuments, in the works of Aristotle and Lucretius, and in the writings of the modern atomic scientist Niels Bohr. Arago analyzed some twenty reports of fireballs in 1838. Some scientists have maintained that they are merely optical illusions, and so Tesla himself thought until they began to appear accidentally on his high-voltage equipment in Colorado.

These strangely ephemeral objects, unlike regular lightning, move slowly, almost parallel to the ground. They have been known to appear in airplanes in flight, move eerily along the floor of the cabin, and after no more than five seconds vanish. In modern plasma physics the most commonly held theory is that the fireball receives its energy from its surroundings by a naturally created electromagnetic field, and that the diameter of the plasma sphere depends upon the frequency of the external field, so that a resonance occurs. But the returns are still not in, and scientists continue to differ (see chapter 30).

Nevertheless, Tesla's speculations do accord with some hypotheses. He thought, for example, that the initial energy was insufficient to maintain the fireball and that there must be another source, which he believed came from other lightning passing through the fireball's nucleus. For him fireballs were merely a fascinating nuisance, yet he took the time to follow this apparently useless research wherever it might lead—and in the process claimed that he had learned how to create the phenomenon at will.[15] Modern scientists, using the most powerful nuclear accelerators, have tried and failed to replicate his achievement (although the fascinating, and potentially valuable, nuisance still occurs unasked).

Another of Tesla's claimed discoveries at Colorado Springs came late one night as he was working at his powerful and sensitive radio receiver. Only elderly Mr. Dozier, the carpenter, remained on duty. Suddenly the inventor became aware of strange rhythmic sounds on the receiver. He could think of no possible explanation for such a regular pattern, unless it were an effort being made to communicate with Earth by living creatures on another planet. Venus or Mars he supposed to be the more likely sources. No one at that time had ever heard of such phenomena as regular sounds from space.[16]

Thrilled and awestruck, he could only sit and listen. Soon he became obsessed with the idea of returning the signal: There must be a way.

The probable explanation of what he had heard was radio waves from the stars. Not until the 1920s were such counting codes again picked up by astronomers (and given official credence); and in the thirties they began to be transmitted as coded numbers into a digital recorder. Nowadays "listening" to the stars is commonplace.

Although Tesla could not doubt the testimony of his ears, he could nevertheless anticipate the ridicule of his fellow scientists when they heard the news. He was therefore slow to reveal his discovery. And the reaction when it came was all that he might have guessed.

Professor Holden, the former director of the University of California's Lick Observatory, was the quickest to criticize: "Mr. Nikola Tesla has announced that he is confident that certain disturbances of his apparatus are electrical signals received from a source beyond the earth," he told a reporter. "They do not come from the sun, he says; hence they must be of planetary origin, he thinks; probably from Mars, he guesses. It is the rule of a sound philosophizing to examine all probable causes for an unexplained phenomenon before invoking improbable ones. Every experimenter will say that it is almost certain that Mr. Tesla has made an error, and the disturbances in question come from currents in our air or in the earth. How can anyone possibly know that unexplained currents do not come from the sun? The physics of the sun is all but unknown as yet. At any rate, why call the currents 'planetary' if one is not quite certain? Why fasten the disturbances of Mr. Tesla's instrument on Mars? Are there no comets that will serve the purpose? May not the instruments have been disturbed by the Great Bear of the Milky Way or the Zodiacal light? There is always a possibility that great discoveries in Mars and elsewhere are at hand. The triumph of the scientists of the past century are still striking proof, but there is always a strong probability that new phenomena are explicable by old laws. Until Mr. Tesla has shown his apparatus to other experimenters and convinced them as well as himself, it may safely be taken for granted that his signals do not come from Mars."[17]

But the last thing Tesla intended to do just then was to disclose his apparatus to other scientists. His work in Colorado was finished. The New Year, 1900, arrived and went almost unnoticed by the inventor, who was in the midst of preparations to dismantle his equipment and depart.

Tesla, at least, seemed perfectly satisfied with what he had achieved in Colorado. He had made lightning dance at his command; he had used the whole Earth as a piece of laboratory equipment; and he had received messages from the stars. Now he was in a hurry to get on with the future.

15. MAGNIFICENT AND DOOMED

When he reached New York in mid-January 1900, reporters and magazine editors pounced upon him.

Predictably, the eastern scientific fraternity had echoed Professor Holden in denouncing Tesla's claim to have received a message of extraterrestrial origin—at least, without telling them how he did it. But Tesla's offense was greater than that. The signals, as he had written to Julian Hawthorne of the Philadelphia *North American* just before leaving Colorado, indicated to him that "intelligent beings on a neighboring planet" must be scientifically more advanced than Earthlings, a suggestion not easily swallowed by doctors of philosophy.

Tesla burned to reply to these "messages" from outer space. Certain that he was at the forefront of a broad, revolutionary technology, he immediately began filing new patents for radio and the transmission of energy, based on his Colorado experiments.

As a first step, he envisioned building a world radio center offering all the services enjoyed today—interconnected radio-telephone networks, synchronized time signals, stock-market bulletins, pocket receivers, private communications, radio news service. He referred to it as a world system of intelligence transmission.

The first patent that he filed on his return (No. 685,012) was a means for increasing the intensity of electrical oscillations, the medium for doing so being liquefied air to cool the coil and thus reduce its electrical resistance. He also received two other patents in 1900 and 1901 related to buried power transmission lines and the method of insulating them by freezing a surrounding dielectric medium such as water. One, a reissued patent (No. 11,865), referred to a "gaseous" cooling agent—apparently a key word that had been inadvertently omitted from his original patent No. 655,838. He was therefore one of the originators of cryogenic engineering.

Many years later, in the 1970s, developmental projects were initiated in America, Russia, and Europe for methods of using superconductors to transmit underground bulk electrical power, employing various cryogenic enclosures. Brookhaven National Laboratory at Upton, New York, has been at the forefront of this international effort. Brookhaven's method resembles Tesla's except that the object of the modern work has been to cool the *conductor* to a few degrees above absolute zero. The similarity moves closer, however, when considering Tesla's 1901 patent No. 685,012, in which he describes the supercooling of conductors to appreciably lower their resistance, thereby minimizing their dissipation when conducting current. This is yet another instance in which his pioneer work has gone unacknowledged—possibly because it might open a door for the U.S. Patent Office to invalidate later claims.

The race to be first with long-range radio transmission appeared to favor Marconi, of whose success the world press had made much during Tesla's absence. Tesla scorned the paltry efforts being made in America, such as signaling the results of yacht races on Long Island Sound. He announced a plan to operate his robotic boat by radio control at the Paris Exposition—from his office in Manhattan!

Meanwhile, as George Scherff reminded him, there was a problem of some urgency with respect to his bank account. He had run through $100,000 during eight months in Colorado.

To whom should he turn? Colonel JJA? George Westinghouse? Thomas Fortune Ryan? J. Pierpont Morgan? C. Jordan Mott? Although he was being ridiculed in the press, his reputation among capitalists still remained good. One thing that impressed such hard-headed gentlemen was the record of the Westinghouse Company in maintaining its monopoly of alternating-current patents despite the efforts of competing industrialists to batter down the walls.

In a search for new developmental capital he again began frequenting the Players' Club in Gramercy Park, the Palm Room of the Waldorf-Astoria, and, of course, Delmonico's. To the same purpose he suggested to a willing Robert Johnson that an article for *Century* magazine be written by Tesla on energy sources and the technology of the future. He slaved over this article, which was eventually entitled "The Problem of Increasing Human Energy," and which appeared in June 1900. Like most of Tesla's writing it turned out to be a lengthy philosophical treatise rather than the brisk report on his Colorado research which Johnson had desired. Nevertheless, it created a sensation.

Part of this was due to the accompanying photographs—some of the many taken in Colorado—in which the inventor resorted to a mild form of trickery involving not just time exposures but double exposures. They depicted him quietly seated on a wooden chair, absorbed in his notes, while enough lightning to kill a roomful of people slashed and snapped around his head. (Even though local photographers were available in Colorado, he had imported a Mr. Alley, a favorite from Manhattan, to record the experiments made with his magnifying transmitter.) The time exposures ran one or two hours in length, which of course produced much denser and more dramatic lightning effects than shots of single discharges would have done. And although the occupant of the chair did not simultaneously sit there—for he would certainly have been electrocuted—Tesla knew that the human focus was needed to heighten the dramatic effect.

It had been a painful modeling job, for the experiments and hence the photographs had to be made at night when the weather was usually below zero. He explained in his diary how it was done: "Of course, the discharge was not playing when the experimenter was photographed, as might be imagined! The streamers were first impressed upon the plate in dark or feeble light, then the experimenter placed himself on the chair and an exposure to arc light was made and, finally, to bring out the features and other detail, a small flash powder was set off."[1] Thus the structure of the empty chair in later exposures did not show through Tesla's body as if in some weird kind of X-ray photo.

The results were as felicitous as even he might have wished for. Everyone who saw these photographs was astounded. When he sent a print to Professor A. Slaby, who was beginning to be known as the Father of German Radio, the latter replied that Tesla must have discovered something unique; he himself had never seen anything like it.

The inventor's Colorado diary discloses that one reason for his constant experimenting with photographs there was his disappointment with the pictorial results of his ball-lightning research. Of this he wrote: "A very important matter is to use better means of photographing the streamers exhibiting these phenomena. Much more sensitive plates ought to be prepared and experimented with. The coloring of the films might also be helpful in leading up to some valuable observations."[2]

He also was thinking further of "the value of powerfully excited vacuum tubes for the purpose of photography. Ultimately, by perfecting the apparatus and selecting properly the gas in the tube, we must make the photographer independent of sunlight and enable him to repeat his operations under exactly the same conditions . . . such tubes will enable him to regulate the conditions and adjust the light effects at will."[3]

The *Century* article with its photographs and predictions thrust him even further into the center of controversy. But though his scientific colleagues sniped, the press remained generally loyal.

"The press at large has of late been having a good deal of fun with Nikola Tesla and his predictions of what is to be done in the future by means of electricity," wrote the Pittsburgh *Dispatch* (February 23, 1901) from Westinghouse country. "Some of his sanguine conceptions, including the transmission of signals to Mars, have evoked the opinion that it would be better for Mr. Tesla to predict less and do more in the line of performance.

"Nevertheless a recent decision in the U.S. Circuit Court for the Southern District of Ohio fully recalls the fact that Tesla is by no means without his record of complete and thorough achievement. . . .

"Mr. Tesla has a wealth of enthusiasm and fertility of imagination with regard to the future that naturally evokes witticisms. But anyone is ignorant of the recent history of electricity who does not know that Tesla stands in the front rank of electrical inventors, by what he has actually accomplished."

From the electrical-engineering editor Thomas Commerford Martin came eloquent support: "Mr. Tesla has been held a visionary, deceived by the flash of casual shooting stars; but the growing conviction of his professional brethren is that because he saw farther, he saw first the low lights flickering on tangible new continents of science. . . ."

Publicity—whether good or bad—was precisely what Tesla wanted, for he still desperately needed to attract the attention of potential backers. One of the first (though not necessarily most important) to step forward was Stanford White, the celebrated architect. The two men met one evening at the Players' Club, which White had just remodeled, and, feeling an immediate rapport, soon fell into an intense conversation. White had read and been excited by Tesla's vision of the future as painted for *Century* magazine. When the inventor began to describe the physical plant that he envisioned for his world broadcasting system, the architect became an eager partner in the grand plan.

Nor was this grand plan a mere fantasy. Even while Tesla had still been in Colorado, oscillators and other equipment were being assembled in his New York shop under the close direction of Scherff and an engineering assistant. Security was tight as usual. Immediately upon his return he got in touch with George Westinghouse, knowing that his engineers could supply the custom-built machinery he would need.

His Colorado experiments, he wrote Westinghouse, absolutely demonstrated the practicability of establishing telegraphic communication to any point on the globe "by the help of the machinery I have perfected." He would need an engine and a direct-current dynamo of at least 300 horsepower on either side of the Atlantic, and these would be expensive.

"You will know of course," he confided, "that I contemplate the establishment of such a communication merely as the first step to further and more important work, namely that of transmitting power. But as the latter will be an undertaking on a much larger and more expensive scale, I am compelled to first demonstrate such feature to get the confidence of capital. . . ."[4] He also requested that Westinghouse lend him $6,000, to be guaranteed by his English royalties.

The industrialist invited Tesla to ride with him on the train from New York to Pittsburgh in his private "palace car" to talk over the whole matter. On the trip Tesla explained that his plant would surpass in performance the Atlantic cable both in speed and in the ability to send many simultaneous messages. He proposed that Westinghouse retain ownership of any machinery he furnished and interest himself to a certain extent in the venture. But Westinghouse had learned his lesson in the hard world of finance. He suggested that Tesla explore financing among capitalists who were looking for opportunities to invest excess wealth.

One such prospect to whom Tesla then turned was Henry O. Havemeyer, otherwise known as the Sultan of Sugar because of his impressive monopoly of refineries. Tesla, lavish in gift-giving whether he had money or not, sent a messenger all the way to Newport, Rhode Island, with an expensive cabochon sapphire ring as a wedding present for the Sultan. Alas, his homage was not at once rewarded.

Others to whom he confided his plans for the world system included Astor and Ryan. Although the full extent of Colonel Astor's involvement in the project is unknown, Tesla must have had some success, for when the former's estate was appraised in 1913, it revealed five hundred shares of stock in the Nikola Tesla Company.

The spring of 1900 passed with Tesla in an agony of frustration. He and Robert read with dismay the newspaper advertisement of F. P. Warden & Company, Bankers: "MONEY . . . Marconi certificates will net you from 100 to 1,000% better results than any labor of yours can produce." The stock of the British Marconi Company had first been offered at $3.00; already it was selling at $22.

Tesla, believing that Marconi had infringed his patents, wanted to sue him. His mood was further inflamed by the final lines of the advertisement: "The Marconi system is endorsed by such men as Andrew

Carnegie and Thomas A. Edison, and by the press of the entire world. Edison, Marconi, and Pupin are the Consulting Engineers of the American Company."

There it was—the three of them in cahoots to cheat him of his invention of radio. Tesla wrote to Robert pretending to feel optimism about the damages he might recover in such a suit, saying, "I am delighted to learn from the enclosed advertisement that Andrew Carnegie has such responsibilities. He is a good man to call on for damages. My stocks are on the rise!"[5]

Of all the people who had read Tesla's article in *Century* magazine and been impressed by the boldness of his vision, one fitted the inventor's requirements perfectly: J. Pierpont Morgan.

The two met for a talk about the world system. Tesla instinctively was less forthcoming than he had been with Westinghouse: no need to distract the financier with too much technical information. Instead, he dilated on themes of money and power. He described to Morgan the plan for all wavelength channels to be broadcast from a single station. Thus the financier would have a complete monopoly of radio broadcasting. Where others in the field were thinking only in limited terms of point-to-point transmissions, as in ship-to-shore and transoceanic wireless, Tesla was talking about *broadcasting* to the entire world. Morgan was interested.

Tesla followed up their meeting with a letter on November 26, 1900, describing exactly what he was offering—up to a point. He had already made transmissions over a range of nearly seven hundred miles, he said, and was able to construct plants for telegraphic communication across the Atlantic and, if need be, the Pacific Ocean. He could operate selectively without mutual interference a great number of instruments and could guarantee absolute privacy of messages. He had all the necessary patents, he added, and was free to enter into agreements.

He proposed that his name be identified with any corporation that might be formed, and estimated a cost of $100,000 for building a transatlantic plant and $250,000 for a Pacific plant, with six to eight months to build the former and one year for the latter.[6]

He made no mention to Morgan of the wireless transmission of power, not because he had given up the idea, but for the prudent reason that it would have made some of the banker's existing investments obsolete. In any event Mr. Morgan could not be expected to be enthusiastic about the prospect of beaming electricity to penniless Zulus or Pygmies.

Morgan replied that he would agree to finance Tesla to the extent of $150,000. That, however, he warned, was as far as he would go. Although he advanced only a portion of this sum and although the country was in the throes of rampaging inflation, which caused Tesla's bankroll to begin shrinking immediately, the latter was nevertheless ecstatic.

The relationship (no doubt a familiar one for Morgan) quickly became like that between courtier and king. Morgan was "a great and generous man." Tesla's work would "proclaim loudly your name to the world. You will soon see that not only am I capable of appreciating deeply the nobility of your action but also of making your primary philanthropic investment worth 100 times the sum you have put at my disposal in such a magnanimous, princely way. . . ."[7]

Morgan, who had no interest in philanthropic business arrangements, responded by sending Tesla a draft of their agreement and asking him to sign over 51 percent interest in his various radio patents as security for the loan.[8]

Tesla sent Morgan a note in which he quoted an admiring comment from Professor Slaby, now a German privy councillor in addition to being a renowned scientist: "'I am devoting myself since sometime to investigations in wireless telegraphy which you have first founded in such a clear and precise manner. . . . It will interest you, as father of this telegraphy, to know. . . .'" This would indicate to Morgan the speciousness of claims being advanced by Marconi and others. Tesla also observed to his patron that neither Raphael nor Columbus could have succeeded without their wealthy sponsors.

With financing apparently assured, Tesla now set about acquiring land on which to build his transmitter. James D. Warden, manager and director of the Suffolk County Land Company, who owned two thousand acres on Long Island, made two hundred acres at Shoreham available to the inventor.[9] The parcel, isolated and wooded, was adjacent to the farms of Jemima Randall and George Hegeman, and sixty-five miles from Brooklyn. The delighted Tesla christened the site Wardenclyffe and visualized it as becoming one of the first industrial parks. Two thousand persons would be employed at the world broadcasting station while their families resided in the surrounding development.

In March 1901, Tesla went to Pittsburgh to place orders with Westinghouse for generators and transformers. At the same time he had agents in England scouting the coastline for a suitable location on that side of the ocean. He was now far too busy to think of the Paris Exposition, which came and went without a world-shaking demonstration by the inventor.

W. D. Crow, an architectural associate of White's, worked closely with Tesla on the design of a tower, which would have at its peak a giant doughnut-shaped copper electrode 100 feet in diameter. Later this was changed to resemble the cap of a gigantic mushroom. The octagonal tower, made entirely of wooden beams preassembled on the ground, would rise from a large brick building. But the total height of this fantastic structure posed a worrisome question because of wind resistance.[10]

On September 13 Tesla wrote Stanford White: "I have not been half as dumbfounded by the news of the shooting of the President [McKinley was shot on September 6] as I have by the estimates submitted by you, which, together with your kind letter of yesterday, I received last night.

"One thing is certain: we cannot build that tower as outlined.

"I cannot tell you how sorry I am, for my calculations show, that with such a structure I could reach across the Pacific. . . ."[11]

For a time they considered falling back on an older design utilizing two, or perhaps three, much smaller towers, but eventually a single tower was built that soared to a height of 187 feet. Within it was a deep steel shaft that ran 120 feet down into the earth. This shaft, encased by a timber-lined well twelve feet square and encircled by a spiral stairway, was

*Nikola Tesla in 1885, aged 29. The portrait is by Sarony,
Tesla's favorite photographer. (Smithsonian Institution,
National Museum of American History)*

Tesla's birthplace. His father's church stands nearby. (Tesla Museum, Belgrade, Yugoslavia, photo by Professor P. S. Callahan)

One of Tesla's original two-phase induction motors. When linked to his polyphase method of generating and transmitting electricity, this motor became the foundation stone on which the modern electrical power industry was built. (Smithsonian Institution, National Museum of American History)

"To Tesla." An autographed picture of Edison, Tesla's first American employer. (Tesla Museum)

George Westinghouse founded the Westinghouse Electric and Manufacturing Co. in 1886 and purchased Tesla's alternating current patents. (Westinghouse Electric Corp.)

Tesla's famous 1891 lecture before the American Institute of Electrical Engineers at Columbia College. (Electrical World)

Some Tesla inventions shown at the Columbian Exhibition. In the center is the induction motor; on the right, a spinning metal egg that demonstrated the effect of alternating current. On the left is the Tesla turbine engine which came later. (Tesla Museum)

A model of a radio-controlled robot ship designed by Tesla in the mid-1890s. Another of these "teleautomatons" was submersible. (Century magazine, 1900)

Tesla in his laboratory, 1898. The device shown is an unconnected coil illustrating the action of two resonating circuits of different frequencies—today one of the basic circuits used in computers. The pressure at the end of the coil facing the viewer (illuminated by streamers) is approximately one half million volts. (Courtesy L. Anderson)

An early version of the inventor's famous steam turbine. Without blades, vanes, or valves, it was operated by the adhesive action of steam spiraling between closely set metal discs. (Tesla Museum)

The Columbian Exhibition at the Chicago World's Fair, 1893, exemplified the new "Age of Light," which Tesla did so much to bring about. (Smithsonian Institution, National Museum of American History)

A photomontage from The World Today *illustrates Tesla's theory that the earth itself could be "split open like an apple" by applying the principle of mechanical resonance.*

Mark Twain and actor Joseph Jefferson in Tesla's laboratory, 1899. (Columbia University Libraries)

A classic Sarony portrait of Tesla in 1894, contemplating his wireless light.
(Tesla Museum)

Discharge of several million volts cascading around Tesla in his Colorado Springs laboratory. The roar that accompanied such discharges could be heard 10 miles away.
(Burndy Library)

Katharine Johnson, from her husband's
book Remembered Yesterdays.
(Courtesy Little Brown and Co.)

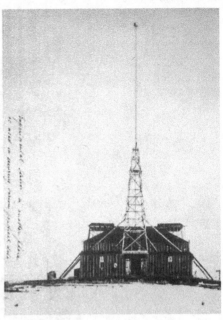

Tesla's Colorado Springs experimental
station. A 30-inch metal sphere tops the
145-foot mast. (Smithsonian Institution,
National Museum of American History)

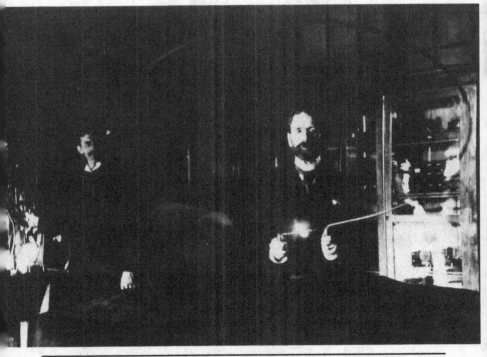

Robert Underwood Johnson with Tesla in the inventor's laboratory.
(Columbia University Libraries)

The ill-fated Wardenclyffe tower built in 1901–03. It was intended for radio broadcasting and wireless transmission of power across the Atlantic. (Courtesy L. Anderson, after photo by Lillian McChesney)

An artist's rendering of Tesla's concept of war of the future. The tower-like structures (based on the intended final form of the Wardenclyffe tower) are directing remote-controlled defenses against robot attackers. As Hugo Gernsback wrote in Science and Invention, where this illustration appeared, "machines only will meet in mortal combat. It will be a veritable was of Science." (Gernsback Publications, Inc.)

The letterhead of Tesla's business stationery recalls some of his more
important inventions. In the center is the Wardenclyffe tower as it was
intended to look when finished.
(Courtesy L. Anderson)

Tesla in his offices at 8 West 40th St., across the street from
the New York City Public Library.
(Columbia University Libraries)

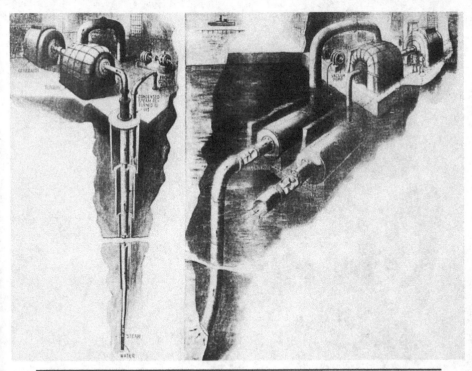

At age 75 Tesla presented two more plans for deriving energy from nature. At left, an earth-heat system; at right, an ancestor of modern ocean thermal energy conversion systems. The sketches are from a 1931 issue of Everyday Science and Mechanics. *(Gernsback Publications, Inc.)*

Some mementos of the days when the dapper Tesla was a well-known man-about-Manhattan. *(Tesla Museum)*

The Tesla bust by Ivan Meštrović, ca. 1939. (Columbia University Libraries)

designed so that air pressure could raise it to touch the tower's top platform. Wardenclyffe was a landmark as magnificent in concept and execution as America's Golden Age of electrical engineering ever produced. Magnificent and doomed.

Because of the inventor's impatience for delivery of his machinery, Westinghouse assigned a special person to expedite it. But the slowness with which Tesla was getting his money from Morgan forced the inventor to take on other work while awaiting the completion of Wardenclyffe. He moved his offices to New York's Metropolitan Tower for increased professional visibility.

One of his schemes for raising money involved developing a special kind of induction motor built by Westinghouse, but there were continuing problems with it. He also installed Westinghouse equipment at the New York Edison plant. Meanwhile, George Scherff went exploring for business opportunities as far away as Mexico.

A great disappointment to Tesla was the government's continuing failure to order his radio-controlled devices for coastal defense. When Congress passed a Coast Defense and Fortification Bill providing $7.5 million, he wrote Johnson that perhaps half a million would be "invested in Teleautomatons of your friend Nikola," and that the rest undoubtedly would find its way into the "hands and pockets of the politicians." Even this note of cynicism betrayed unwarranted optimism.

He soon had cause for bitterness. As the year 1901 drew to a close the world press blazoned the news that Marconi, on December 12, had signaled the letter "S" across the Atlantic Ocean from Cornwall to Newfoundland. What astonished Morgan and many others was that he had done it without anything like the great plant that Tesla was building.

They doubtless did not know that Marconi had utilized Tesla's fundamental radio patent No. 645,576 filed in 1897 and issued March 20, 1900. Small wonder that Tesla began to refer bitterly to the "Borgia-Medici methods" by which he was being deprived of credit and fortune. But radio technology was then a mystery to most scientists, let alone the average investment banker.

Angry though he was, Tesla wasted no time on sour grapes but kept his eyes on the magnificent obsession rising from the farming land of Long Island. At first he nursed it along from a private home near the construction site. When Scherff moved out from Manhattan to expedite the work, Tesla returned to his stylish retreat at the Waldorf-Astoria to keep a finger on the pulse of Wall Street. Each day he and Scherff exchanged several wires and letters. And since Wardenclyffe was only an hour and a half from New York by train, at least once a week the inventor, elegantly attired down to his gray spats and accompanied by a Serbian manservant bearing an immense hamper of food, entrained for Long Island.

He worried constantly about security. Across the Sound residents of New Haven watched in fascination as the octagonal tower rose like a mushroom grower's fantasy above the tree line of the North Shore. As for the townspeople in nearby Shoreham, they believed themselves to be on the brink of fame and industrial prosperity.

16. RIDICULED, CONDEMNED, COMBATTED

As the "wonder tower" lifted its airy spars ever higher, Tesla drove himself and a large staff without mercy. He sent money to Germany for radio engineer Fritz Lowenstein's return, and the latter soon joined the Wardenclyffe team. Another well-known engineer, H. Otis Pond, who had worked for Edison, helped build the laboratory.

Years later Pond was to say that he disagreed with history's assessment of the two inventors. Edison was "the greatest experimenter and researcher this country has produced—but I wouldn't rate him as much of an originator," he said. Tesla, however, he considered "the greatest inventive genius of all time."[1]

Pond often accompanied Tesla on long walks. They were together on the day in December 1901 when Marconi sent the first transatlantic signal. "Looks like Marconi got the jump on you," he said.

"Marconi is a good fellow," replied Tesla. "Let him continue. He is using seventeen of my patents."

Pond also recalled Tesla's worrying about the instruments of war that he had been inventing. He had just launched his model wireless torpedoes in the Sound, encircled a ship with them, and landed them on the beach. "Otis," he said, "sometimes I feel that I have not the right to do these things."[2]

The inventor's hectic schedule often gave the impression that he was three or four individuals. His New York laboratory had become a meeting place for scientists from all over the world. The nights were filled with social activities, arduous experimental work, the writing of patent applications, professional-journal articles, and letters to editors.

Seeing and being seen by the "right" people compelled him to function as both a day and night person; nights in a row passed during which he scarcely closed his eyes. An inevitable consequence of this frenetic schedule was that his friends became compartmentalized, occupying cells of his life that others were unaware of. Intimates such as the Johnsons, for example, had no idea of the prominence or even the identity of some of his newer confidants, which is not to say that they were ever displaced in his affections.

The daylight hours were important for beseeching his patron, Morgan, to advance funds more rapidly; for reminding him that inflation was threatening to sink the ship. He met with other potential investors. He pleaded with manufacturers to expedite machinery and advance credit. And while he remained in New York, he wrote daily letters of instruction to Scherff.

One welcome event in this hectic year of 1902 was a visit to the United States by England's famous Lord Kelvin, who proclaimed himself in complete agreement with Tesla on two controversial issues: 1. that Mars was signaling America; and 2. that the conservation of nonrenewable resources was of critical importance to the world.[3] Kelvin, like Tesla, was convinced that wind and solar power should be developed to help save coal, oil, and wood. Windmills, he declared, should be placed on roofs at the earliest opportunity, to run elevators, pump water, cool houses, and heat them in winter.[4]

Edison, however, differed with his distinguished contemporaries, putting off the evil day of shortages for "more than fifty thousand years." The forests of South America alone, he argued, would provide fuel for that long.

When Kelvin expressed high praise for the "scientific prophets" of America, it was an obvious appreciation of Tesla and came as balm for the inventor's spirit. After a banquet in Kelvin's honor at Delmonico's, the Englishman proclaimed that New York was the "most marvelously lighted city in the world," and the only spot on Earth visible to Martians.

Perhaps inspired by the excellent wine, he declared, "Mars is signaling . . . to New York." The announcement made headlines in all the next day's papers. When Tesla had made a similar assertion, the air had been filled with controversy. Now that a man of Kelvin's stature had said it, not a single demurrer was raised by the scientific community, even including Professor Holden. This sudden change of attitude inspired Tesla's friend Hawthorne to write a misguided article that went farther than Kelvin's sensational announcement. Obviously, he wrote, the men of Mars and of other older planets had been visiting the Earth and looking it over year after year, only to report back, "They're not ready for us yet." However, once Nikola Tesla had been born, things had changed. "Possibly they (the starry men) guide his development; who can tell?"[5]

This one line alone may indict the romantic Hawthorne for planting the seed later nourished by those who would adopt Tesla as their pet Venusian and in so doing harm his scientific reputation.

Thus, Hawthorne went on, it had been the inventor in the lonely observatory on the mountain-flank for whom the first message was gently rapped out. "Another might have heard it and neglected it. . . . But Tesla, whose brain, compared with those of most of his contemporary scientists, is as the dome of Saint Peter to pepper-pots, had been trained to the hour, and the signal was not in vain."[6]

Although no one ever accused Nikola Tesla of lacking ego, we may imagine that he had to grit his teeth when he sat down to his writing desk to thank his friend for this embarrassing literary flight. "That was very nice," he wrote, "all except the dome of St. Peter's and the pepper-pots!"

Then he prudently changed the subject and went on to speak of his scientific concerns: "Half the time I am like a man condemned to death and half the time the happiest of mortals. All is still but *hope*. It may take centuries but I feel it in every fibre it is coming sure! One thing is settled in my Colorado experiments. We can construct a machine which will carry a signal to our nearest neighbors as certainly as across your muddy Skykoll [sic] river. We can also feel safe about receiving a message, provided there are other fellows in the Solar System knowing as well as we know how to operate this kind of apparatus. . . ."[7]

In June Tesla moved his laboratory from Manhattan out to the new brick building at Wardenclyffe. Here, except for the exigencies flowing

from the project itself, the demands on his time would be fewer. Only
workers were admitted to the grounds. The isolation and quiet were just
what he needed.

What with one thing and another, when he was summoned for jury
duty on a murder case in New York that fall, he put the notice aside and
forgot it. Soon, to his embarrassment, newspaper headlines made him
sharply aware of an American citizen's duties: "Nikola Tesla Fined
$100—Fails To Show Up for Jury Duty in General Session—Is Sorry
Now." So he was, and reported at once to court where he apologized. He
was then excused from duty on the ground of his opposition to the death
penalty. *The New York Times* quoted him as saying that capital punishment was "barbarous, inhuman, and unnecessary."[8]

Marconi remained the hero of the hour in America as elsewhere.
Tesla's doings, by comparison, seemed merely mysterious. In February
1903 the *Electrical Age* carried a critical article about "Nikola Tesla—His
Work & Unfulfilled Promises." Wrote the author: "Ten years ago Tesla
was the electrician of greatest promise. Today his name provokes a regret
that a promise should have been unfulfilled." It had been too long since he
had scored a clear triumph, and he was learning how short mortal memories could be.

By spring (1903) Tesla's money problems had grown so severe that
he was again compelled to return to New York to try to raise more funds.
Even so, he did not entirely put aside his scientific preoccupations. In a
note to Scherff, one of hundreds, he asked that there be sent to Professor
Barker of the University of Pennsylvania "the photograph (Roentgen) of
the bones of a hand . . . taken in Colorado . . . the tube was operated without wire by my system. . . ."[9]

When he returned to Long Island, it was for the raising of the fifty-
five-ton, sixty-eight-foot dome frame onto the top of the tower. (The
plans had called for covering the dome with copper plates to form an
insulated ball, but this was never done.) Scherff took the occasion to
remind him that funds were dangerously low. Creditors were impatient.
Even when Morgan sent the remainder of the promised $150,000, it had

scarcely covered outstanding bills. And Tesla felt that Morgan, with his great power over the national economy, had been responsible to a large degree for the rising costs.

He wrote the financier on April 8: "You have raised great waves in the industrial world and some have struck my little boat. Prices have gone up in consequence twice, perhaps three times higher than they were...." [10]

Morgan, his capital still heavily committed to railroad centralization and other sensible enterprise, declined to advance more funds. Two weeks later Tesla again wrote: "You have extended me a noble help at a time when Edison, Marconi, Pupin, Fleming, and many others openly ridiculed my undertaking and declared its success impossible...."

But Morgan still did not act, and Tesla, now beginning to feel pangs of desperation decided to play his final card. So at last he wrote to Morgan and bared his true goal—not *just* the sending of radio signals but the wireless transmission of power.

On July 3, he wrote: "If I would have told you such as this before, you would have fired me out of your office.... Will you help me or let my great work—almost complete—go to pots?...." [11]

The answer came eleven days later, addressed to N. Tesla, Esq.: "I have received your letter," wrote Morgan, "... and in reply would say that I should not feel disposed at present to make any further advances." [12]

Tesla replied that night in Jovian style by going to the tower and setting off such a fireworks display as no one had seen before. His tests went on through the night and for several thereafter. Residents watched in awe as blinding streaks shot off from the spherical dome, at times lighting up the sky within a radius of hundreds of miles. Take that, Pierpont Morgan, they seemed to say.

When reporters rushed to the scene they were turned away. The New York *Sun* reported, "Tesla's Flashes Startling, But He Won't Tell What He Is Trying For at Wardenclyffe. Natives hereabouts ... are intensely interested in the nightly electrical display shown from the tall tower where

Nikola Tesla is conducting his experiments in wireless telegraphy and telephony. All sorts of lightning were flashed from the tall tower and poles last night [July 15]. For a time, the air was filled with blinding streaks of electricity which seemed to shoot off into the darkness on some mysterious errand. When interviewed, Tesla said, 'The people about there, had they been awake instead of asleep, at other times would have seen even stranger things. Some day, but not at this time, I shall make an announcement of something that I never once dreamed of.'"

Even stranger things? Was it simply a journalistic tease?

In Colorado he had achieved voltages of ten to twelve million volts on the antenna sphere of his magnifying transmitter, although he believed that 100 million volts were feasible. On his return to New York he applied for another group of patents of which the most important was an "Apparatus for Transmitting Electrical Energy," related to the Wardenclyffe project, being No. 1,119,732 filed in 1902 but not issued until 1914.* In fact it was filed only weeks after Marconi's transatlantic wireless success.

The problems of getting investment capital for unfinished Wardenclyffe were compounded in the fall of 1903, when the so-called Rich Man's Panic struck. Now, the chances of wooing Morgan back into the fold seemed more remote than ever.

Aided by his loyal friends, Tesla redoubled his efforts to raise money. Lieutenant Hobson pulled all his strings trying to interest the Navy in robotry. Having seen Tesla's radio-controlled boats and torpedoes in 1898, he urged him to display them in a naval exhibit at Buffalo and set it up so that the inventor would not have "the usual difficulties of formalities." But in vain.

The naval hero reported that there had been a fight within the Navy about Tesla's wireless exhibits—a feud not directly related to his inven-

* First in this series was No. 685,012, means for increasing the intensity of electrical oscillations by use of liquefied air, issued in 1901; followed by 655,838, method of insulating electrical conductors; 787,412, art of transmitting electrical energy through the natural medium; 723,188, method of signaling; 725,605, system of signaling; 685,957, apparatus for utilization of radiant energy; and 1,119,732.

tions, he said, but rather something ongoing between two high officers, resulting in the rejection of Tesla's entry.[13] It is possible that Hobson was skewing the situation to avoid hurting his friend's feelings.

Tesla then went to Thomas Fortune Ryan and succeeded in raising a little supplemental funding. But it all went into paying off existing creditors, whose bills were beginning to tower like Wardenclyffe itself. He did not need patient, observant George Scherff to tell him where the trouble lay. "My enemies have been so successful in representing me as a poet and visionary," he said, "that it is absolutely imperative for me to put out something commercial without delay."

In the years ahead he would repeatedly stagger forth from avalanches of debt to strike out anew on some practical scheme for commercializing his inventions. Whether he was less fortunate as an independent than his old foe Edison, it would have been hard to say, but certainly their lives were following different paths.

Edison in his late fifties was wealthy but ill with a variety of afflictions, including mysterious lumps in his stomach that had appeared during his X-ray research (they finally vanished). Disappointed in his ore ventures, increasingly deaf, he had withdrawn from emotional contact with family and friends. He had gone into semiretirement, was old before his time, and not only could afford, but felt obliged, to hire a full-time bodyguard for himself and his household. Such were the stigmata of success.

There was a growing interest in the medical profession in Tesla's therapeutic oscillator, a small Tesla coil. Doctors and professors phoned from all over the country, saying they were constantly receiving inquiries for such high-frequency apparatus. Scherff told Tesla he could easily start a thriving business in medical apparatus, with a crew of thirty men and an investment of $25,000. He predicted a rapid profit of $125,000, almost as much as Morgan's total investment in Wardenclyffe.

The inventor told him to go ahead with such work at Wardenclyffe but did not himself seem much interested. Instead he issued two handsome brochures, one describing the world system of communication and another, expensively printed on vellum, that announced his entry into the field of consulting engineering.

The main work crew were kept busy fabricating and assembling novel devices, blowing glass vacuum tubes, and doing the routine work of firing up the steam generator. The latter job was spasmodic: by mid-July of 1903 the paying of coal bills had become a problem. Periodically the crew was laid off.

When coal for the Wardenclyffe generator could be afforded, the inventor wired Scherff to stoke up for a weekend of tests and took a train to Long Island. "The troubles and dangers are at their height," he wrote Scherff on one occasion. "Coal problem still awaits solution. The Wardenclyffe specters are hounding me day and night. . . . When will it end?"[14]

Scherff, now moonlighting as a bookkeeper for other firms, lent small sums of money when he could. Dorothy F. Skerritt later verified a report that he probably lent the inventor a total of $40,000 over the years.[15] "Tesla seemed to have Mr. Scherff hypnotized," she observed.

In an earlier and better time, as the inventor told her, he had been able to get money from Morgan just by asking for it. On one occasion the financier signed a blank check and told Tesla to fill in what he needed. Tesla said that the amount was $30,000. But now Morgan's disenchantment with Wardenclyffe was final. Tesla, equally firm in his determination to forge on, sent more letters—at first persuasive and beseeching, then angry, accusing, and bitter. They pursued the banker by special messenger everywhere, even to the pier as he embarked on yet another grand tour of the Continent.

Inevitably rumors spread that Morgan had acquired Tesla's radio patents just to prevent their development; but there was no proof. When bad news whispered along Wall Street, it gained strength from itself. Word that Morgan was dropping out of the world system venture—he actually had been only a lender—convinced other potential backers that it must be a soap bubble.

Tesla knew such rumors were killing him; but there was little he could do except live each day trying to dodge bill collectors, pleading with other bankers and rich acquaintances, working out scientific problems of the project, seeking to market other inventions, and bidding for consultancies.

The multiplier effect of his hard luck knew no geographic limits. He was sued for nonpayment of electricity furnished to the experiment station at Colorado Springs, and this was odd, considering that Leonard Curtis, one of the owners of the City Power Company, had assured him that electricity would be free. The city of Colorado Springs also sued him, for water bills. Finally the caretaker of his old experiment station brought an action for unpaid balance of wages due him.

Tesla's response to the city was Teslian. Inasmuch as he had graced it with his presence and had erected his famous station there, he wrote, he believed it should feel privileged to pay for the water.

He ordered the old lumber from the station sold and the money therefrom paid to the power company. And finally he returned to Colorado Springs to appear in court with his attorney to answer the caretaker's suit. The plaintiff was awarded a judgment of about $1,000. A sheriff's sale of laboratory fixtures paid some of it. The rest Tesla was to pay in $30 increments dragged out over half-a-dozen years.

Then for a time it seemed as if his luck was turning. Money began trickling in from the sale of medical coils, which were now being manufactured on an assembly line at Wardenclyffe for hospitals and research laboratories. And he managed to invent a new turbine of revolutionary design, which he felt sure would restore his fortune and reputation.

Although partying with his friends continued, there was a new frenetic quality about it, as if the celebrants had begun to sense the tragedies ahead and were determined to lose no opportunity for laughter.

Katharine sent invitations to come and meet the usual parade of celebrities, and reproaches when he failed to do so. One note ended typically: "We shall soon be far away but then you would never know it. You do not need anybody, inhuman that you are. How strange it is that we cannot do without *you*."[16]

She and Robert were preparing for another sojourn in Europe. Robert's dilettantism was unflagging. "Mrs. Johnson tells me you are going to dine with the Countess of Warwick," he wrote Tesla. "Will you be good enough to ask her grace if the Warwick vase could have been the original of Keats' 'Ode on a Grecian Urn'?"[17]

Significantly, Johnson had begun to worry about whether his holding of stock in the Nikola Tesla Company might be misconstrued by his employers, in view of the several articles and consultations Tesla had been asked to provide to *Century* magazine. He suggested to the inventor that the money he had invested ought better to be construed as a loan, with the stock as security. Such concern about conflict of interest implied that Tesla's stock as a scientist was slipping and that his name no longer carried the old cachet.

Many in the business world appear to have believed that Tesla was still receiving "princely" royalties from Westinghouse on his alternating-current patents, not realizing that he had been bought out at bargain rates in 1896. This was made clear by an article in the Brooklyn *Eagle* of May 15, 1905, calling attention to the "expiration" of Tesla's valuable patents. The newspaper reported that "a great stir" had been created among electricians by announcement that the patents had expired: "There will be a grand scramble everywhere to make the Tesla motor now universally used without paying any more royalty to Tesla. The Westinghouses announced they have a number of subsidiary patents, and will fight."

For it to have become known that Tesla was receiving *nothing at all* would have cast him in a strange light, not to his credit in the world of nonpoets.

Late on the night of July 18, 1905, he wrote to Scherff, anxious at not having heard from him. "The last few days and nights have been simply horrible," he confided, referring to an unnamed illness. "I wish I were at Wardenclyffe in a patch of onions and radishes. *Troubles are at their height.* As soon as things are ready I will come out. We must get much better results."

Only days later he wrote of worries about materials and of taking measures to prevent "the *kind* of accidents we have had before.

"I will tell you frankly that it looks blue for this week unless L. carries out his promise. . . . I have several chances and many hopes but I have been deceived so often that I am a pessimist."

He had been experimenting—for what purpose is unknown—with extremely high-pressure jets of water, of about 10,000 pounds per square inch. A tiny jet, if struck by an iron bar, would deflect it exactly as if another bar had hit it. Such streams of liquid power had destructive effects on any metal with which they came in contact. One day the cast-iron cap of the pressure cylinder broke, and a large fragment shot past Tesla's face to tear a hole in the roof.

On another day Scherff had his face seared while pouring hot lead into screw holes in the flooring. The lead struck water that had been used earlier to swab the floor and exploded upward. Tesla, a few feet away, was only slightly injured but Scherff was seriously burned. For a while it was feared that he would lose his vision.

Considering the dangerous equipment with which the men worked daily, however, the accidents were remarkably few.

Hobson, now a busy young naval recruiting officer, traveling all over the country yet sandwiching in campaigns for political office as well, never failed to call on Tesla when in Manhattan. He fretted about the inventor damaging his health with strain and overwork, tried to lure him to football games at the Naval Academy, and wrote encouragingly from his parents' home in Greensboro, Alabama. "My father and mother say they would rather meet you than any man in the world and that you must come down and visit us and rest up from your herculean tasks. . . ."

He wrote from Texas, from trains, from hotels all over the country, and often from the Army & Navy Club in New York. He spoke of the disappointments of his political candidacy; but in 1903 he finally resigned from the Navy to make a successful career of politics.

Then on May 1, 1905, "the hero" wrote Tesla of "the greatest happiness that has come to me"—his forthcoming marriage to Miss Grizelda Houston Hull of Tuxedo Park, New York.

"Do you know, my dear Tesla," he wrote, "you are the very first person, outside of my family, that I thought of. . . . I wish to feel you present, standing close to me, on this occasion so full of meaning in my life. . . . You occupy one of the deepest chambers of my heart. . . ."[18]

Two years later Hobson succeeded in winning election to Congress from his home state of Alabama and would serve there until 1915. He would distinguish himself—to Tesla's dismay—by becoming a leader of the prohibition movement. The inventor considered alcohol in reasonable amounts an ambrosia, although his devotion to the naval hero would survive such ideological differences.

Mark Twain, seventy years of age and relishing his fame, returned to America. He and Tesla sought each other's company as often as their work and other demands permitted, meeting usually at the Players'.

Katharine, distressed by Tesla's commuting to Long Island, scarcely knew from one day to the next where to send her invitations. "I will be here this evening," she wrote, "but suppose you are throughout the week at your country residence in the remote wilds of Long Island. However, if you happen to be rusticating at your favorite resort, the Waldorf, send me a line when you receive this and let me know when I may expect you. . . . I want to see if you have grown younger, more fashionable, more proud. But whatever you may be you will always find me the same."[19]

This invitation was unusual in its use of the singular pronoun; Robert apparently was traveling, or he was otherwise unable to entertain. Almost certainly Tesla did not accept.

But the early winter brought them all together for a holiday celebration on Thanksgiving eve. Tesla's thank-you note to Katharine urged her not to despise millionaires since he was still hard at work trying to become one. "My stocks have gone up considerably today," he wrote. "If it continues for a few weeks like this, the globe will be girdled soon."[20]

Katharine sent another appeal urging him "to come for my sake as I need cheer and who is all potent as you. . . ." He put her off.

Christmas he normally spent with the Filipovs. She wrote five days before the Yule to remind him, adding, "You must come here tomorrow evening as I want to see you for many reasons, to know how you are. But why try to enumerate them? You know them all except one. I have something to tell you by way of Germany.... When I wrote you last Sunday morning I sent you my first thoughts out of sleep. I knew that you were depressed but did not know why. Please let me have a word *dear* Mr. Tesla that I may have something to count on, something to expect...."[21]

The winter passed with his anxieties over Wardenclyffe mounting daily until it seemed there would be no end to his trials.

The long steaming summer returned to New York, with Tesla's routine seldom varying. To Scherff he wrote again of money problems: "Troubles and troubles, but they do just seem to track me. The Port Jefferson Bank will have to get along with interest, assuming that I can scratch it together."

Soon afterward, however, he hastened to send exciting news. He had had a meeting with Mr. Frick, the industrialist and nouveau-riche art collector. Since becoming manager of the Carnegie Steel Company trust in the 1890s, he had managed to double the size of the plant through the assiduous use of exploited labor and cheap materials. Now, enjoying the rewards of his prescience, he was casting about for new investments. The inventor's note to Scherff exuded optimism: "Troubles are many but progress is encouraging. Had a *very promising* session with Mr. Frick and am full of hope he will *advance capital still necessary.*"[22]

At about this time Tesla and Johnson had an editorial exchange over Hertzian waves. Tesla had sent the latter an article for *Century* that puzzled Johnson because of his assertion that such waves were not employed in wireless telegraphy.

"There is Hertzian telegraphy in *theory* only," Tesla explained, "since these waves diminish very rapidly with distance." Hertz and Crookes, he said, did not apply sources of power since they used the Ruhmkorff coil and a simple spark gap. Tesla claimed that he had made no progress in the field until he was inspired to invent his oscillation transformer, with which he obtained greatly magnified intensity. He believed, after experimenting with different forms of aerials, that the signals picked up by the instruments were actually induced by earth currents instead of being etheric space waves.

Kenneth Swezey, however, later wrote, "Tesla understood well the nature of Hertzian waves and constantly used them. His obstinacy in refusing to admit that these waves played a significant part in the operation of his wireless power equipment . . . merely helped confuse judges and lose cases for him throughout his lifetime."[23]

After his "very promising" session with Frick, the inventor was again forced to send bad news to Scherff: the negotations had come to nothing.

The year 1906 threatened to be, if possible, worse than its predecessor. Even his friend Westinghouse seemed to be avoiding him. Tesla's need for Westinghouse machinery at Wardenclyffe remained almost as urgent as his need for capital. Thus he wrote to the industrialist asking, "Has anything happened to mar the cordiality of our relations? I would be very sorry, not only because of my admiration for you but for other serious reasons.

"The transmission of power without wires will very soon create an industrial revolution and such as the world has never seen before. Who is to be more helpful in this great development, and who will derive from it greater benefits than yourself?"[24] Westinghouse, although knowing that without Tesla's AC patents his firm could not have become the lusty adolescent that it was, replied in effect thanks but no thanks.

The harrowing routine continued. Scherff wrote that a promised carload of coal had not yet been delivered and that scheduled tests must be delayed. He also mentioned tactfully his extra job for two days a month keeping books for a sulfur manufacturing firm. This was a bad omen for Tesla, for Scherff would soon become a full-time employee of this company.

Worse news lay ahead. On June 26, 1906, the newspapers were filled with sensational accounts of the murder of Stanford White. The architect had been shot three times by a Pittsburgh financier, Harry K. Thaw, the night before on the roof of Madison Square Garden, while numerous members of the New York "400" looked on. White was believed by the killer to have been involved with his wife, Evelyn Nesbit, in a love triangle. Later Thaw was committed to the Matteawan Hospital for the Criminally Insane.

But the architect who had given New Yorkers such splendid edifices as Madison Square Presbyterian Church, the Garden City Hotel, the Hall of Fame at New York University, and the Astor Mansion at Rhinebeck, was gone—leaving the tower on Long Island as his final monument.

Scherff left Wardenclyffe that fall. He never ceased, however, to keep an eye on Tesla's financial affairs, working for him on evenings and weekends and almost always remembering to file his tax returns on time.

The world system for broadcasting—a concept designed to incorporate almost every aspect of modern communications—was all over but the mourning. Yet as long as the tower stood, Tesla continued his efforts to complete it.

Exactly when all the workers left, no one could say. Thomas R. Bayles, the general passenger agent of the railroad station just across the road from the abandoned plant, only noticed that passengers had stopped getting off there. A caretaker remained on duty for a time. When curious journalists or research engineers showed up they were allowed to climb to the tower top with its sweeping view of Long Island Sound. For all that the tower looked so light, it was built entirely without metal, even down to the wooden pegs holding together the wooden uprights and cross members. After abandoning the plan for covering the dome with a copper sheathing, Tesla had installed a removable disk through which a beam of radiation could be projected to the zenith.

The visitors found the laboratory filled with curiously complex apparatus. In addition to much glass-blowing equipment there were a complete machine shop with eight lathes, X-ray devices, a great variety of high-frequency Tesla coils, one of his original radio-controlled robot boats, and exhibit cases filled with thousand of bulbs and tubes. There were an office, library, instrument room, electrical generators and transformers, and great stocks of wire and cable.[25] But after the watchman left, vandals entered, broke things, ransacked files, emptied paper on the floor and trampled them.

"It is not too much to say," wrote a Brooklyn *Eagle* reporter, "that the place has often been viewed in the same light as the people of a few centuries ago viewed the dens of the alchemists or the still more ancient wells of the sorcerers. An atmosphere of mystery hung over the place, an unearthly influence seemed to be radiated from the alembic . . . as if drawn down from interstellar space and spread over the countryside to inspire wonder and awe in the minds of the nearby farmers and villagers. . . ."[26]

In 1912 a judgment of $23,500 against the inventor for machinery supplied to the project was won by Westinghouse, Church, Kerr & Co. The equipment left at the site was taken to satisfy it.

Tesla, in order to maintain his fashionable mode of life at the Waldorf through the years, had given two mortgages on Wardenclyffe to the hotel's proprietor, George C. Boldt. They secured bills of about $20,000. He had asked that the mortgages not be recorded, fearing damage to his financial credibility. When in 1915 he was at last unable to make any payments at all, however, he signed the Wardenclyffe deed over to Waldorf-Astoria, Inc.[27]

The hotel corporation tried to convert its strange security into cash but no one in those days knew what to do with the ruins of a world broadcasting center. The War Department was approached for ideas but nothing came of it. Next it was considered as a site for a pickle factory. Tesla must have wept when he heard this. But nothing caught on. And in 1917, rumors began to circulate that German spies were holed up in the magnificent tower, spying upon Allied shipping and radioing signals to U-boats. On July 4, 1917, an explosion of dynamite was discharged inside the tower. Newspapers and even the *Literary Digest* reported that it had been blow up by the U.S. government to halt espionage.[28] Tesla denied the rumor.

In fact the tower was destroyed under a salvage contract between the owners and the Smiley Steel Company of New York, but the inventor did not wish to disclose the real owners. And it was destroyed only in an effort to realize a few dollars from scrap.

The tower proved to be more strongly built than its destroyers guessed. They had to keep blasting away as if it were rooted to the spot by some mysterious force. On the ensuing Labor Day it collapsed, dynamite having triumphed at last over the merely celestial. It brought the corporation $1,750 above salvage costs. A junkman noticed some of Tesla's notes blowing down the street.

"I did not exactly cry when I saw my place after so long an interval," the inventor wrote to Scherff, "but I came very close."[29]

Marconi, with Carl F. Braun of Germany, won the Nobel Prize in physics in 1909 for their "separate but parallel development of the wireless telegraph."

Never for the rest of his life would Tesla give up on his concepts of power transmission and broadcasting. It was not a dream, he declared, "but *a simple feat of scientific electrical engineering*, only expensive—blind, fainthearted, doubting world."

Humanity, he wrote, was not yet sufficiently advanced to be willingly led by "the discoverer's keen searching sense." But perhaps it was better "in this present world of ours that a revolutionary idea or invention instead of being helped and patted, be hampered and ill-treated in its adolescence—by want of means, by selfish interest, pedantry, stupidity, and ignorance; that it be attacked and stifled; that it pass through bitter trials and tribulations, through the strife of commercial existence. So do we get our light. So all that was great in the past was ridiculed, condemned, combatted, suppressed—only to emerge all the more powerfully, all the more triumphantly from the struggle."[30]

Next to Tesla and society, the greatest loser when Wardenclyffe fell was Morgan. There can be little question that he could have written his own ticket for an early lead in radio broadcasting, with a station operating on several adjacent-frequency channels, transmitting in multiplex mode, and thereby *far* surpassing the performance of the slow, single-channel transatlantic cable. Among the many who would use Tesla's patents in the development of commercial radio (legally or illegally), one firm would soon be sending messages a distance of 9,000 miles. The clarity of Tesla's understanding of radio should not be confused with his efforts to transmit electricity wirelessly. He did not confuse them.

17. THE GREAT RADIO CONTROVERSY

E rrors once committed to print are stubborn. With respect to the invention of radio, they have permeated many reference sources, histories of science, scientific biographies, and popular journals. The confusion—partly caused by Tesla himself—was officially cleared up in 1943 when the U.S. Supreme Court reversed an initial finding in Marconi's favor to rule that Tesla had anticipated all other contenders with his fundamental radio patents.*

The radio-engineering fraternity made a major effort to atone in 1956, on the occasion of Tesla's one hundredth anniversary. It is strange therefore to find in the *Dictionary of American Biography* an article on Tesla by an eminent professor of electrical engineering and computer sciences, who cites reference sources through the forties, fifties, and sixties—yet completely omits reference to the U.S. Supreme Court's landmark decision.[1] Even more curious, this author cites articles on Tesla by Anderson, O'Neill, Swezey, and Haradan Pratt (*Proceedings of the Institute of Radio Engineers*), each of whom had done careful research to set the record straight. Popular historians both in the United States and Europe have consistently repeated the error.

Although the news has yet to penetrate encyclopedias, modern radio-engineering authorities now accord Tesla clear priority in a field that for years was confused by seesawing claims involving such international luminaries, besides Marconi, as Lodge, Pupin, Edison, Fessenden, Popov, Slaby, Braun, Thomson, and Stone, to name only the more famous of the pioneers.

Dr. James R. Wait writes: "The simple picture shown based on

* June 21, 1943—"United States Reports; Cases Adjudged in the Supreme Court of the United States," Vol. 320; *Marconi Wireless Telegraph Co. of America v. United States,* pp. 1–80.

Tesla's disclosure in 1893 is the birth of wireless communication. Admittedly, it follows the erudite theoretical and experimental investigations of Hertz who demonstrated the action at a distance from a spark gap discharge. But, by a few years, it precedes Marconi's inventions and practical demonstrations of wireless telegraphy."[2]

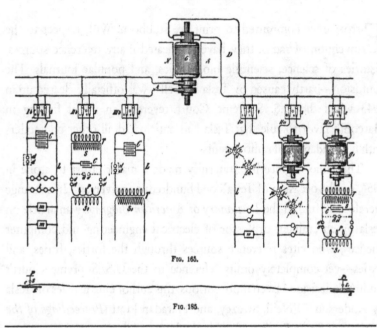

Fig. 165.

Fig. 185.

Figures 165 and 185, referred to in the United States Supreme Court case, are from Tesla's 1893 lecture and are frequently cited as evidence supporting his claim of invention of radio.

Anderson points out that some have confused the argument with respect to the principles of transmission and reception of radio signals with the matter of transmitting voice—an important improvement made practical by DeForest's Audion, or triode vacuum tube. "In a discussion of priority in the invention of radio, one must be very specific about definitions," he writes. "In the . . . case of the Marconi Wireless Telegraph Company of America vs. United States (which was decided June 21, 1943, against the Marconi Company and striking down the fundamental

Marconi patent), the following definition evolved out of the exhaustive depositions taken from many technical experts in the fields of radio and the physical sciences:

"'A radio communication system requires two tuned circuits each at the transmitter and receiver, all four tuned to the same frequency.' The definition does *not* embrace variable modulation that DeForest's Audion provided and through which the transmission and reception of voice and music was made possible. It does *not* address the mode of electromagnetic propagation—that is, ground wave and/or sky wave and the effect of the former on the latter. It does, however, implicitly describe the deliberate, selective transmission at a specific frequency and the selectable reception at that same frequency."[3]

Marconi's original patent application was filed on November 10, 1900, and was rejected on the prior art disclosed by Sir Oliver Lodge. Tesla's first patent was granted in 1898. Moreover Tesla was specific as to which of his patents were directed to wireless power transmission as opposed to signal communication, although this appears to have confused some of the critics of his radio patents.

The U.S. Supreme Court found that Tesla's patent No. 645,576, applied for September 2, 1897, and allowed March 20, 1900, anticipated the four-circuit tuned combination of Marconi.[4]

Tesla, long before anyone else, published in *Electrical World and Engineer* (March 5, 1904) what has always stood out as the clearest statement by any pioneer working in the wireless art *of what radio was to become and as we know it today*. He envisioned the entire concept of transmission of intelligence, not just the sending of a single message from one point to another—and he alone of the pioneers in radio did so.

Tesla said his "World Telegraphy constitutes, I believe, in its principle of operation, means employed and capacities of application, a radical and fruitful departure from what has been done heretofore. I have no doubt that it will prove very efficient in enlightening the masses, particularly in still uncivilized countries and less accessible regions, and that it will add materially to general safety, comfort, and convenience, and maintenance of peaceful relations. It involves the employment of a number of plants, all of which are capable of transmitting individualized signals to the uttermost confines of the earth. Each of them will be preferably located near some important center of civilization, and the news it receives through any channel will be flashed to all points of the globe. A cheap and simple device, which might be carried in one's pocket may then be set up anywhere on sea or land, and it will record the world's news or such special messages as may be intended for it. Thus the entire earth will be converted into a huge brain, capable of response in every one of its parts. Since a single plant of but one hundred horse-power can operate hundreds of millions of instruments, the system will have a virtually infinite working capacity, and it must needs immensely facilitate and cheapen the transmission of intelligence."

These ideas were also discussed by him in *Century* magazine in June 1900 following his return from Colorado.

Another pioneer of radio, J. S. Stone, said in reviewing a field that included Lodge, Marconi, and Thomson: "Among all those, the name of Nikola Tesla stands out most prominently. Tesla with his almost preternatural insight into alternating current phenomena that enabled him . . . to revolutionize the art of electric-power transmission through the invention of the rotary field motor, knew how to make resonance serve, not merely the role of a microscope, to make visible the electric oscillations, as Hertz had done, but he made it serve the role of a stereopticon. . . . [I]t has been difficult to make any but unimportant improvements in the art of radio telegraphy without traveling, part of the way at least, along a trail blazed by this pioneer who, though eminently ingenious, practical and successful in the apparatus he devised and constructed, was so far ahead of his time that the best of us then mistook him for a dreamer."[5]

Among the many authorities in radio who seconded this view (although perhaps not quite soon enough for justice's sake) was Gen. T. O. Mauborgne, former head of the Signal Corps and chief signal officer of the U.S. Army. In *Radio-Electronics* (February 1943, only weeks after Tesla's death), he wrote: "Tesla 'the wizard' . . . captured the imagination of my generation with his flights of fancy into the unknown realms of space and electricity . . . [saw] with astounding vision far beyond his contemporaries, very few of whom realized until many years after the work of Marconi that the great Tesla was the first to work out not only the principles of electric tuning or resonance, but actually designed a system of wireless transmission of intelligence in the year 1893."[6]

Even Professor Pupin of Columbia University, testifying as an expert witness for the Atlantic Communication Company in a suit for alleged infringement of patent brought by the Marconi Wireless Company of America (Pupin's side as expert witness was to change somewhat with time and circumstances), stated on May 12, 1910:

"When William Marconi was 'a mere strip of a lad working for Signor Riggie' in Italy he grounded both wires out of curiosity in an experiment to see what would result and he produced wireless waves without ever fully realizing the full significance of it." But Pupin gave the credit for discovering wireless to Nikola Tesla, who "gave his discovery free to the world."[7]

Another radio-engineering pioneer, Cdr. E. J. Quinby, USN (Ret.), has recalled from his personal experiences in the early days of commercial radio development in America:

"While others fought bitter word-battles in our courts over whose patents were really valid on the all-important system of tuning to avoid wholesale radio interference, nobody seemed to recall that Tesla had covered the subject back before the turn of the century with his comprehensive and fundamental patent on tuning of electrical circuits to resonance. Without this feature, today's ever-expanding radio service would be utter chaos. A Supreme Court decision was finally reached in 1943, crediting Tesla with having anticipated all the others, thus making subsequent patents on the subject null and void."

Tesla himself failed to accomplish his dream of a world wireless system, Quinby pointed out, but he lived to see this all done by utilization of the system he so clearly outlined.

"The high-frequency alternators Tesla built between 1890 and 1895 produced up to 20 kHz, despite the critics who said it couldn't be done, and who accused him of being an impractical dreamer," wrote Quinby. "It remained for Prof. Reginald A. Fessenden to demonstrate that such machines could produce the required quiet carrier for voice modulation, thus eliminating the background roar of the damped-wave spark and arc transmitters with which others were experimenting. Fessenden agreed with Tesla, that the damped-wave transmitters were an abomination, and that the future successful radio development rested on continuous-wave generators."[8]

Thus on Christmas Eve, 1906, and New Year's, 1907, Fessenden startled and delighted listeners up and down the East Coast of the United States and precipitated a flood of fan mail by broadcasting voice and music programs from his transmitter at Brant Rock, Massachusetts. He was using a high-frequency alternator which he built, based on Tesla's design and principle.

During World War I, says Quinby, with the engineering talents of Steinmetz, Alexanderson, and Dempster, the General Electric Company at Schenectady succeeded in scaling up the small experimental models of radio-frequency alternators into the giant 200 kw production model, the first of which was installed at the Marconi Worldwide Wireless Station at New Brunswick, New Jersey, to replace the unsatisfactory high-power spark transmitter.

Ironically, Tesla was among dignitaries invited to witness the inauguration of reliable transatlantic service at this station. President Woodrow Wilson's Armistice terms were carried by radio from the station to Kaiser Wilhelm in April 1919.

Commander Quinby adds: "Later, when President Wilson made his historic voyage to Europe aboard the S.S. *President Washington*, voice communication was established between the New Brunswick station and President Wilson while he was at sea—thanks to the pioneering of Nikola Tesla in demonstrating his high-frequency alternator back in 1895."

Galling as it was to Tesla, however, it remained undeniably true that

Marconi first riveted the attention of the world with his successes in radio and thereafter cleverly held the lead in development with the Marconi Worldwide Wireless Company.

On May 13, 1915, Professor Pupin testified yet again as an expert witness for the defendant in a suit brought by Marconi against the Atlantic Communication Company. This time he appeared to suggest that he himself had invented the wireless "before either Marconi or Nikola Tesla had discovered it," according to press reports of the trial.[9]

In his own experiments, he said, he had found a wireless wave but had not realized its importance. Tesla, he iterated, however, "had given his discoveries to mankind, and this is one of the points on which the Atlantic Company experts expect to deny the claims of Marconi to certain wireless patents."[10]

Tesla himself sued Marconi at last, in August 1915. The Marconi Wireless Telegraph Company of America also sued the U.S. government for allegedly infringing on "Marconi's" patents during World War I. The war of the wireless patents was waged back and forth for decades, and little wonder that confusion ensued.

A thorough account is contained in Anderson's "Priority of Invention of Radio—Tesla vs. Marconi," a monograph for *The Antique Wireless Association* (New Series) No. 4, March 1980. Anderson reports that radio pioneer Major Armstrong added an interesting—if somewhat confusing—sidelight to the controversy. He wrote to Anderson shortly before his own death in 1953, saying that in his opinion Tesla was the true inventor of the guided weapon (robotry) but that there had been efforts to reduce his claim to that invention. Moreover, said Armstrong, he did not believe Tesla should be advanced as the inventor of radio.

"By his writings on the problem of signaling without wires, he fascinated and inspired some of the early workers in the field, Marconi perhaps himself included," wrote this protégé of Pupin who was now himself a famous inventor in radio.

"However . . . he failed to conceive or to experimentally discover that vital idea uncovered by Marconi which brought practical wireless signaling into being. I have pointed out if he had gone ahead on the basis of his erroneous theory, he would have been very likely to have discovered the principle that Marconi did uncover and so would have become known as the inventor of wireless telegraphy. But this he failed to do and so the credit quite properly goes to Marconi."[11]

Tesla's fame, Armstrong went on, was "secure on the basis of his accomplishments in the power field, and as a prophet of the possibility of wireless, and of wireless-controlled engines of war."

Armstrong almost appeared to be saying that because Tesla was famous as an explorer of one continent of science or two, it was unimportant that he be given due credit in a major third field. Perhaps in part this curious view reflected the growing academic commitment to specialization: generalists were out of style, hence the existence of any lurking Leonardo ought to be denied.

Armstrong offered to confide to Anderson the "vital secret" that made Marconi's work a success and Tesla's a failure. In January 1954 Anderson asked for this. He was distressed shortly afterward to learn of Major Armstrong's unexpected death. But later, he said, two scientists who knew Armstrong and of his "unswerving championing of Marconi" told him that Armstrong was referring to the ground connection in a transmission-reception system. Anderson was stunned.

"Every one of Tesla's patents for either communication or power transmission showed the ground connection," he wrote me. "In fact, the matter of ground conduction was the cornerstone of Tesla's concept. Nonetheless, and despite the fact that *the* Marconi patent was declared invalid by the Supreme Court, Armstrong stuck to his position. I guess that's what confounded me so much—Armstrong's overlooking an essential point already so clearly in evidence."[12]

Haraden Pratt, fellow of the Institute of Radio Engineers* and past chairman of the IRE History Committee, has written that Tesla's radio ideas and the apparatus he produced were left for others to pick up and embody for less ambitious but more practical purposes.

* The IRE is now incorporated in the Institute of Electrical and Electronics Engineers, Inc., or IEEE.

"For this reason," he noted, "Tesla's influence on the development of radio was known to but a limited number of people. A few eminent persons who attended or read his lectures during the 1890 decade were inspired by his revelations and some others, who later delved into the background of the art, became aware of the pioneering import of his contributions.

"Far ahead of his time, mistaken as a dreamer by his contemporaries, Tesla stands out as not only a great inventor but, particularly in the field of radio, as the great teacher. His early uncanny insight into alternating-current phenomena enabled him, perhaps more than any other, to create by his widespread lectures and demonstrations an intelligent understanding of them, and inspired others not yet acquainted with this almost unknown field of learning, exciting their interest in making improvements and practical applications." [13]

In sum, it is easier to see in retrospect than during Tesla's time how the truth came to be obscured.

Ample rewards went to those scientists, inventors, and engineers who successfully got in on the ground floor of commercial radio. Tesla, spending more time in his ivory tower than on ground floors, was to be smiled on fitfully by fame and in the long run ignored by fortune.

In his later years an incident occurred that revealed the true depths of his feelings about the great radio controversy. On a day in January 1927 a young Yugoslav named Dragislav L. Petković, visiting America, arranged to call upon him. He then lived on the fifteenth floor of the Pennsylvania Hotel at 34th and Broadway. Times were hard and he had grown reclusive. Petković was invited to have lunch in his rooms and treated to a spread of California fruits and vegetables, fish, and honey.

After conversing for a time, Petković tried to learn the secret of the animosity between Tesla and Pupin. Once he had asked Dr. Pupin about this matter. The latter had burst out, "How long will our people celebrate only mysterious persons, instead of what's clear to everyone to understand?"

Now, when he put the same question to Tesla, the inventor frowned and raised his hand as if to protect himself from something very unpleasant. After a pause, he explained to Petković that in the early days in America, when he and Pupin were both struggling to survive, the latter had asked him for help with the English language. According to Tesla, he was having difficulty holding a job with the telephone company. Tesla helped Pupin but later somewhat tactlessly reminded him of the favor. Pupin angrily said that he himself had been quite able to do the work and that Tesla "did not do anything for him." Tesla was hurt but forgot the matter.

Later, however, when he lectured at Columbia College, demonstrating his transformer and his theories of radio and electrical-power transmission, "Mr. Pupin and his friends interrupted my lecture by whistling, and I had difficulty quieting down the misled audience." But this was not the worst.

"During the lawsuit which I've instituted against Mr. Marconi for stealing my apparatus and drawings from the Patent Office," Tesla allegedly continued, "Mr. Pupin, called to testify on my behalf as a countryman, went on the side of Mr. Marconi, who, after three years of legal battle was forced to admit under oath that the transmission of power to long distances is my invention."

Tesla paused, then added, "Let the future tell the truth and evaluate each one according to his work and accomplishments. The present is theirs, the future, for which I really worked, is mine."

With tears in his eyes but with a smile, he resumed his meal. He and the visitor quietly attacked their cantaloupe. Then the visitor asked another question.

"Can you tell me something about Mr. Marconi?"

It was one of the few recorded occasions when Tesla ever lapsed from courtesy. He laid down his spoon.

"Mr. Marconi," he said, "is a donkey."[14]

18. MIDSTREAM PERILS

The inventor, now aged fifty, his reputation as a scientist under serious attack, had seldom looked more debonair. He was still slim, smooth-faced, and young-looking, his hair as thick and black as ever. He still dressed like a fashion plate, had a wide circle of friends, and clung however tenuously to his cherished residence at the Waldorf-Astoria Hotel.

Indeed, Tesla's relationship with the hotel may have come as close to a marriage as anything he ever experienced. A life not lived regally seemed to him scarcely worthwhile. Always one to confront disappointment with panache, he seemed to have a special talent for floating elegantly through the worst of times. It was not that he never worried about debts but simply that his mind, preoccupied with ideas, could screen them out over long periods. Thus he could chide less fortunate worriers like Scherff and Johnson for alleged faintheartedness in the face of financial adversity. Yet the psychic importance of money to the inventor, apart from his real need for it, seemed to grow in contrast to its declining accessibility, as his many letters to Johnson, Scherff, and others made clear.

Although in appearance and in mode of life Tesla was to go on much as before, inwardly he had begun to change. His bitter disappointments in the early years of the century exerted a corrosive and lasting effect upon his personality. He wrote revealingly to George Westinghouse on the latter's latest enforced corporate reorganization to say, "the strength of a man shows itself in adversity." Unfortunately, adversity also tends to reveal weaknesses.

Tesla became an inveterate writer of self-serving letters to newspapers. Where in palmier years he had been generous in praising the achievements of both his predecessors and his contemporaries, and had seldom troubled to reply to personal critics, he now became prickly and shrill in self-defense. He was quick to put down competitors, the weak as well as the powerful, and to claim priority of discovery on his own behalf. Cheated too often, he grew even more secretive in the protection of his patents. The psychic damage to him had been real and deep.

Tesla was fortunate in the early years of the century to attract two loyal, intelligent women to his staff as secretaries, both of whom went on in later years to important careers of their own. Both, needless to say, had trim figures.

Muriel Arbus was a charming blonde who assisted Tesla with patent claims and, after his death, went on to distinguish herself as the head of Arbus Machine Tool Sales in New York—the only woman in America at the time to have created her own firm as a buyer of large machine tools. She was extremely successful.

Dorothy Skerritt joined Tesla in 1912, witnessed many demonstrations at his laboratory at 8 West 40th Street, and often went across the street to the New York Public Library to do research for the inventor. A person who met both women observed that Skerritt "seemed to be more aware of the underlying motivations of individuals and sensed the implications of adverse circumstances, yet said little. Arbus, on the other hand, took things at face value and seemed to enjoy talking about them."

Skerritt had worked for a patent-attorney group before joining Tesla and remained with him until 1922. Arbus spent World War II working for the Office of Production Management, the War Production Board, and later for the Reconstruction Finance Corporation, after which she began her own unusual business.

As for their mutual employer, more and more in the years ahead Tesla would advance scientific claims recklessly, discussing them with reporters fresh from the moment of inspiration without subjecting his ideas either to experimental verification or even much reflection. At times he would seem almost megalomaniacal. Some journalists, interested only in headlines and bylines, quoted him without question, but those who cared for him, like O'Neill and Swezey, sought to save him when necessary from his own announcements.

Edison had merely reflected the sniping of the professors when he had taunted, "Tesla is a man who is always going to do something." But presumably such a charge could have been made against Edison himself by anyone choosing to overlook the sweep of his solid achievement in favor of his unrealized aspirations. He too courted reporters, inveterately promising more than he could deliver.

Professor Joseph S. Ames of the Johns Hopkins University had written an early attack on Tesla all too typical of the view from academe, a comparison of the works of Marconi, Pupin, and Tesla in which the latter came off a miserable third: "The Tesla motor, so called, and the electrical machines which are modifications of it, are known to the world, and so is the 'Tesla coil,' which is a simple improvement of one of Henry's instruments; but as yet no discovery bears his [Tesla's] name. ..."[1]

This attack, like others of its ilk, was, of course, simply wrong-headed. By the late 1920s $50 billion would be invested in Tesla's nineteenth-century induction motors and systems of power transmission throughout the world. He was "the father of radio" and of automation. Most universities, including Johns Hopkins, already relied on Tesla coils in their research laboratories. And a whole series of other original inventions had been patented, many of them before 1900, by the man of whom Ames could write "no discovery bears his name."

But it was also true that Tesla was more often an originator of broad concepts than of discrete innovations. His lectures radiated ideas that many others took in hand, applied practically and subsequently patented. Indeed, this was one of the reasons why he was now beginning to play his cards so much closer to his chest.

If at the same time he seemed to sensationalize his new projects and theories, it was because, acting as his own entrepreneur, seeking financial backing from investors and the wealthy, he resorted to methods that would appeal to them. The shows staged in his laboratories were intended to dazzle the money people who, he realized, would not be technically able to "steal" his ideas. Fellow scientists, jealous but not deceived, were naturally unhappy.

Despite the fact that his cornucopia of ideas flowed almost as richly as ever, he had reached an age when he could no longer ignore his own mortality. Friends and acquaintances began to fall away. Mark Twain died in 1910, and the loss affected Tesla deeply. Three years later Morgan also died, as great a pivotal figure in national affairs as he had been in Tesla's own career.

Tesla's psyche had always been a festival of neuroses, but now his behavior seemed to become, if anything, stranger still. No one knew when the inventor began gathering up the sick and wounded pigeons and carrying them back to his hotel. Usually, however, it was a mission that he carried out late in the day.

His whole routine was that of a night person. It was also that of a prince of the blood. To hotel servants he could be cavalier and cutting one moment, generously rewarding with tips the next.

As a night person he arrived at his office promptly at noon; as a prince of the blood he required that Miss Arbus or Miss Skerritt be standing just inside the door to take his hat, cane, and gloves. Then all window shades would be drawn to simulate the darkness in which he worked most productively. In fact, the only time when the shades were raised was when a lightning storm was flashing over the rooftops of the city. Then he would lie upon a black mohair couch to watch the northern or the western sky. His employees said that he had always talked to himself, but that during these lightning storms, when he insisted on being alone, they could hear him through the door and that he became positively eloquent.

But despite all the stresses and anomalous symptoms, Tesla's creative genius remained unimpaired. In 1906, the year of his fiftieth birthday, in the wake of many trials, he built the first model of his marvelous turbine. Possibly it had been inspired by his childhood efforts to build a vacuum motor and by his plans, during the year he spent living in the mountains, for shooting mail through a tube beneath the ocean. Possibly the idea for the bladeless turbine went back even further—to his earliest memory of invention, when he had built a tiny waterwheel that had no blades but spun all the same.

Whatever its provenance, the model weighed less than ten pounds and developed thirty horsepower. He later built much larger ones that developed 200 horsepower. "What I have done," Tesla explained, "is to discard entirely the idea that there must be a solid wall in front of the steam and to apply in a practical way, for the first time, two properties which every physicist knows to be common to all fluids but which have not been utilized. These are adhesion and viscosity."[2]

Julius C. Czito, the son of Tesla's long-time machinist Kolman Czito, built several versions of the turbine in his machine shop at Astoria, Long Island. The rotor of the so-called "derby hat powerhouse" consisted of a stack of very thin disks of German silver, mounted on the center of a shaft. They were enclosed in a casing provided with ports. "When deriving energy from any kind of fluid," Tesla elaborated, "it is admitted at the periphery and escapes at the center; when, on the contrary, the fluid is to be energized, it enters in the center and is expelled at the periphery. In either case it traverses the interstices between the disks in a spiral path, power being derived from or imparted to it, by purely molecular action. In this novel manner the heat energy of steam or explosive mixtures can be transformed with high economy. . . ."[3]

He saw no limits to its uses. With gasoline fuel it could power automobiles and airplanes. It could drive ocean liners across the Atlantic in three days. It could be used for trains, trucks, refrigeration, hydraulic gearing (motion transfer), agriculture, irrigation, and mining—and it would run on steam as well as gasoline. He was even designing a futuristic automobile that he planned to power with it. Above all, he believed that the turbine would be inexpensive to manufacture compared to traditional models.

His spirits were greatly bolstered when the Tesla turbine began to be widely acclaimed—in concept. Even the War Department officers declared it to be "something new in the world," and said they were "greatly impressed with it." It seemed reasonable to expect that a fortune was to be made by the man who had designed a better rotary engine.

Tesla began to emerge from the endless trauma of humiliation and debt. The scalding nightmares were occurring less often in which the death of his brother Daniel so long ago, his mother's death, and the destruction of Wardenclyffe seemed all mixed up. All he needed now was capital, and the turbine would put him back on top. He began ticking off in his mind the names of possible investors.

19. THE NOBEL AFFAIR

The many mourners who crowded into the funeral for J. Pierpont Morgan at St. George's Church in Manhattan on April 14, 1913, were attending a theatrical closing, the end of a long run of history. Tesla had been sent tickets for the gallery, with apologies that better seats were not available.

After the rites the inventor thoughtfully set his calendar ahead exactly one month. On May 14 he asked for an appointment with J. P. Morgan, the scion of the House of Morgan.

The younger banker and the inventor met and talked mainly about the commercial potential of the Tesla turbine. Six days later the inventor received a loan of $15,000 from the J. P. Morgan Company with interest at 6 percent, for nine months.[1]

Tesla followed up their meeting with a letter describing in forceful and fluent words the uniqueness of his latest invention. "Knowing this as I do," he wrote, "not merely as an expert but as a seer, you may judge how anxious I am, for the sake of the world, to connect myself with men of your integrity and power...."[2]

Unfortunately he did not stop there. He could not refrain from reminding Morgan junior that Morgan senior had lent him $150,000 for Wardenclyffe. Others had let him down in this venture, he said; otherwise the first world broadcasting system would by then have been flourishing. Accordingly, he proposed the formation of two new companies, one for the development of radio broadcasting and the other for turbine manufacture, offering "to turn over to you my entire interest in both," leaving it to Morgan to accord to him such a part as he might choose.

The younger Morgan replied stiffly that he could not possibly consent to Tesla's turning over an interest in the two companies. Instead, he suggested that Tesla go ahead and organize the two firms and, from his profits, start repaying the $150,000 to the Morgan estate as and when he could. This did not end the dialogue, but it certainly crimped it.

Over the next several years the inventor favored J. P. Morgan with repeated invitations to invest in a wireless station and the turbine. But the financier neither understood nor was much interested in fluid propulsion or radio. As for the wireless transmission of electrical power, the old objection still obtained: Why would Morgan want to put all his power lines out of business? Nevertheless, the financier lent Tesla $5,000 and then, like his father, took refuge in a European vacation. He sailed that autumn, carrying some books the inventor had given him, and leaving Tesla pacing the dock.

Meanwhile Tesla began licensing his turbine in Europe. Through the intercession of the former Prince Albert of the Belgians, he received $10,000 for the license in Belgium. A concession in Italy was expected to bring him $20,000. In America he concluded automobile and train lighting contracts and was working on other practical arrangements. But still his funds were far short of his needs.

He struggled to take the disappointments philosophically and had a remarkably accurate idea of his own place in—or rather, out of—time.

"We are but cogwheels in the medium of the universe," he wrote to Morgan, "and it is . . . an unavoidable consequence of the laws governing that the pioneer who is far in advance of his age is not understood and must suffer pain and disappointment and be content with the higher reward which is accorded to him by posterity."[3]

When Morgan returned just before Christmas, Tesla presented him with a number of propositions. He was again desperate. "I am almost despairing at the present state of things. I need money badly and I cannot get it in these dreadful times. You are about the only man to whom I can look for help. . . ." He closed by wishing the multimillionaire a Happy Christmas. Morgan responded with a bill for interest of $684.17 on the two loans already extended and a hearty return of seasonal wishes.

In January 1914, despite the threatening World War, Tesla pleaded with Morgan that he needed another $5,000 to finish and ship a turbine to the German Minister of Marine, High Admiral Alfred von Tirpitz. He felt that no question of loyalty was involved since he had already offered the turbine to the U.S. government. Despite complimentary remarks about his invention from some in the War Department, no orders had been placed at home. This time Morgan relented and extended another loan.

Two months later he offered Morgan a chance to finance an automobile speedometer and to buy a two-thirds interest in a new company. It was becoming painfully apparent that there were problems with the turbine: the metal had not been manufactured that could take such high speeds for long; and it was by no means inexpensive, at least in the early stages of development. More time was needed and therefore, he must develop interim sources of capital.

But this time Morgan's secretary returned all the enclosures and advised that Mr. Morgan could not possibly be interested in any further inventions.

Through the next winter, however, Tesla continued to appeal to Morgan again and again. "Please do not take this as another cry for help," he wrote; but in reality it was a desperate cry. Meantime, he moved his offices from the smart Metropolitan Towers to the less expensive Woolworth Building. In November Morgan replied that he would extend the loans but would add nothing more to them.

Everyone seemed to be hard up. Scherff sent the inventor two new notes of his own for signature, replacing the old unpaid ones, so that the former employee might be able to use them as collateral. He expressed disappointment that Tesla had been unable to make at least some payment; but Tesla, signing the new notes, wrote glowingly of his prospects for the turbine.

In the midst of personal trials he still found time to help his friends. Johnson, who had been promoted to editor of *Century* magazine four years earlier, wrote urging secrecy in an office scandal that jeopardized his position. He referred to a letter from a Mr. Anthony "written without any knowledge of the situation at the office. What he will say when I tell him the new situation, the Lord knows. . . ."[4]

Tesla, having interceded in the mysterious affair, wrote back that he had done all he could to dispose of the matter, "but I have encountered resistance and so fear I have reached no tangible results . . . I am not relaxing my efforts. Trusting that you will not let this little embarrassment weigh too heavily on your mind. . . ."[5]*

But the little embarrassment—the nature of which remained a closely guarded secret—resulted in Johnson's resignation. Things were never quite the same at their fashionable home on Lexington Avenue after that. Although in time Robert obtained a new position as permanent secretary of the American Academy of Arts and Letters, his finances seem to have undergone some sort of erosion. The Johnsons continued to indulge themselves with the parties, the servants, and the European holidays to which they were accustomed, but now their lifestyle was bringing them into debt. A pattern began that would continue for the remainder of both men's lives, of borrowing small sums of money from each other to cover overdrafts. More often, surprisingly, it was to be Tesla who bailed out Johnson.

War with Germany was drawing ever closer for the United States. Tesla and young John Hays Hammond, Jr., on the latter's initiative, corresponded at length on possible ways of earning money through military applications of their work in robotry. Hammond, using Tesla's principles, had built an electric dog on wheels that followed him everywhere, its motor operated by a light beam behind its eyes. Bowser was not exactly an invention to set the generals and admirals fiercely bidding against one another, but Hammond had also operated a crewless yacht by radio in Boston harbor, and the two inventors toyed with the idea of forming a teleautomatics company. Hammond had an automatic selec-

*At one time much earlier Tesla had been a close friend of Richard Watson Gilder's, who preceded Johnson as editor in chief of *Century*. From many exchanges of invitations between Tesla and Gilder, followed by Gilder breaking such engagements, it appeared that Mrs. Gilder frowned upon her husband's friendship with the inventor. On January 24, 1898, Tesla wrote to Mrs. Gilder: "We have all been greatly disappointed for not having Mr. Gilder with us . . . apologizing for bothering you. . . ."

tive system he wanted to develop, and Tesla thought that a dirigible torpedo he had invented many years earlier could be of service to the War Department. But although he helped Hammond get a technical article published on the state of the art, their efforts at joint development were not pursued.[6]

Even at this stage in his career Tesla was still often handicapped by public confusion about his citizenship. A Washington *Post* article, making a common error, referred to him as the "noted Balkan scientist." And among the bureaucrats in Washington he may have suffered from a mistaken application of the NIH (not invented here) factor. Mere superiority of product would seldom be enough to override such a disadvantage, however much society might be the loser.

But no doubt much more damaging to Tesla's prospects at this time were those traditional enemies of innovation—inertia and vested interest. An industrial consultant tells of inquiring some years ago of an executive in the Office of Naval Research in Washington, D.C., if they had ever sponsored R&D programs on the Tesla turbine. The reply was: "We get proposals all the time for funding Tesla turbine work. But let's be candid. The Parsons turbine has been around a long time with entire industries built around it and supporting it. If the Tesla turbine isn't an *order of magnitude* superior, then it would be pouring money down the rat hole because the existing industry isn't going to be overturned that easily. . . ."

Sometimes Tesla's inventions had better luck backing into America from abroad. In 1915 a German firm, licensed to use his wireless patents, built a radio station for the U.S. Naval Radio Service on Mystic Island near Tuckerton, New Jersey. It was equipped with the famous Goldschmidt high-frequency alternator of the magnetic reflecting type, which enabled radio frequency alternating currents to be developed directly.[7] Tesla received royalties of around $1,000 per month from these patents for two years—a most welcome source of income.

When the chief engineer, Emil Mayer, told him that messages from the station were being received at a distance of 9,000 miles, he took the news calmly, for it merely confirmed what he already knew. "You have thus proved, practically, what I demonstrated with my wireless plant in scientific experiments carried on from 1899 to 1900," he replied. Unfortunately the war soon brought his radio royalties to a halt. The Tuckerton Radio Station was closed by the government in 1917, the year of America's entry into the war. Tesla did, however, receive royalties later from the Atlantic Communications Companies.

World War I was brought home to America's Serbian population long before it engaged the country at large. Local Slavs could not help but feel the impact since Serbia had led the movement for Pan-Slav unification that ultimately set off the whole conflagration. A Serbian nationalist assassinated the Archduke Francis Ferdinand at Sarajevo in Bosnia, leading to both Serbia and Montenegro being overrun by the Central Powers, composed, among others, of Austria and Germany. Soon news of the extreme sufferings of the Serbian people reached the United States.

Relief efforts were begun by local emigrants under the auspices of the Serbian Orthodox Church and the Serbian Red Cross, of which Pupin was chairman. Further proof of the antipathy between the two scientists is supplied by an anecdote from this period. The very Rev. Peter O. Stijacić with a noted professor of theology from Serbia called upon Tesla one day to solicit a message of unity to American Serbs, in hope of inspiring them to send more generous aid to the homeland. Innocently they suggested that such an appeal be signed by the famous Nikola Tesla, Michael Pupin, and Tesla's dear friend, Dr. Paul Radosavljević (known as Dr. Rado), who taught at New York University. Tesla politely asked that they excuse him as a signer, knowing the impossibility of ever agreeing with Pupin on a word or a phrase, let alone a message of unity. And if the unification committee itself could not get together, . . . American Serbs, he said philosophically, but with amusement in his eyes, had minds of their own.

In 1918 a kingdom of Serbs, Croats, and Slovenes was proclaimed under the rule of King Peter I. But this by no means ended Slavic turmoil and misery. Eleven years later King Peter's successor, Alexander I, following a move toward separatism by Croatia, established a dictatorship.

At least the country then acquired one name for all its people and its parts—*Yugoslavia*. Tesla approved both of Alexander and of unity.

Another anecdote about the inventor is told by the Reverend Stijacić. On his first trip to America as a young writer for the Serbian Federation, Stijacić had been surprised to find in the Chicago Public Library, a book of poems, the author of which was the popular Serbian poet Zmaj-Jovan. The translator was Nikola Tesla. Later, when Stijacić was taken by Dr. Rado to meet the inventor in his offices on the twentieth floor of the Metropolitan Tower, he said, "Mr. Tesla, I did not know that you were interested in poetry."

A look of wry amusement shone in the inventor's eyes. "There are many of us Serbs who sing," he said, "but there is nobody to listen to us."

The New York Times on November 6, 1915, carried a story on page one, based upon a Reuter's dispatch from London, reporting that Tesla and Edison were to share the Nobel Prize in physics. Interviewed the next day, Tesla told a *Times* reporter he had received no official notification of the award. But he speculated that it might be for his discovery of a way to transmit energy without wires. This, he said, had proved to be practical not only over terrestrial distances but "even effects of cosmic magnitude may be created."

He then described for the reporter a future when all wars would be waged with electrical waves instead of explosives. More positively, he said, "We can illuminate the sky and deprive the ocean of its terrors! We can draw unlimited quantities of water from the ocean for irrigation! We can fertilize the soil and draw energy from the sun!"[8]

Asked what he thought Edison was being honored for, Tesla tactfully replied that Edison was worthy of a dozen Nobel prizes. That gentleman, reached in Omaha on his way home from the Panama-Pacific Exposition in San Francisco, seemed surprised when shown the London dispatch. He too said he had received no official notice. He made no further comment.

Robert and Katharine were unsurprised but delighted by the news. The former quickly sent his congratulations. Tesla, his mood now more thoughtful, replied that many people would win the Nobel Prize, but that "I have not less than four dozen of my creations *identified with my name in technical literature*. These are honors real and permanent, which are bestowed, not by a few who are apt to err, but by the whole world which seldom makes a mistake, and for any of these I would give all the Nobel prizes during the next thousand years."[9]

What followed was a curious business. The western press, including leading magazines, picked up the story and without checking, gave it wide circulation. In another story in the *Times*, Tesla was interviewed again as a Nobel winner.

His comments to the inquiring reporter were entirely typical. He bemoaned the fact that the world, after so many years, still did not understand his concepts of voice transmission. With such a plant as Wardenclyffe, he explained, the telephone exchange of New York City could hook up, enabling subscribers to speak to anyone in the world without any change in the telephonic apparatus. A picture from the European battlefields could be transmitted to New York in five minutes.

The current passed through the earth, he elaborated, starting from the transmission station with infinite speed from that region and, slowing down to the speed of light at a distance of 6,000 miles, then increasing in speed from that region and reaching the receiving station with infinite velocity.

"It's a wonderful thing. Wireless is coming to mankind in its full meaning like a hurricane, some of these days. Some day there will be, say, six great wireless telephone stations in the world system connecting all the inhabitants of this earth to one another not only by voice but by sight."[10]

Flawed though his physics might be (Tesla would resist to the end the idea that light sped faster than anything) his prophecy was sound enough. He did not explicitly foresee today's microwave-boosting synchronous satellites for television, yet something of the sort had been in his mind since, as a teenager, he had envisioned building a ring around the equator that would revolve in Earth synchrony.

And if he did not invent television, he at least imagined it. Four years later Johnson suggested that as a money-making venture, Tesla invent a way of reproducing football games on a home screen as they occurred. "I am already expecting to become a multimillionaire without going into show business," he replied, but went on to offer his "best suggestion," which was to employ "nine flying machines, winged and propellerless five hundred miles or more, take negatives, develop films, and reel them off as they arrive. . . . It calls for an invention to which I have devoted twenty years of careful study, which I hope will ultimately realize, that is television, making possible to see at distance through a wire. . . ."[11] But, in fact, he never pursued this idea.

The report of the Nobel Prize in physics for 1915, to be jointly shared by Edison and Tesla, was carried in the *Literary Digest*[12] and *The Electrical World* of New York,[13] both publications having gone to press before November 14, the date on which another Reuter's dispatch, this time from Stockholm, dropped a devastating bombshell. The Nobel Committee announced that the prize for physics would in fact be shared by Professor William Henry Bragg of the University of Leeds, England, and his son, W. L. Bragg of Cambridge University, for their use of X rays to determine the structure of crystals.

What had happened? The Nobel Prize Foundation declined to clarify. One biographer and close friend of Tesla's reported years later that the Serbo-American had declined the honor, stating that as a discoverer he could not share the prize with a mere inventor.[14] Yet another biographer advanced the theory that it was Edison who objected to sharing the prize, that it was in keeping with his "sardonic and sadistic brand of humor" to have deprived Tesla of $20,000 when he knew how much he needed funds.[15]

But no real evidence exists to prove that either of them declined the Nobel. The Nobel Foundation said simply, "Any rumor that a person has not been given a Nobel Prize because he has made known his intention to refuse the award is ridiculous." The recipient would have nothing to say in the matter, except to decline it after the fact if he or she so chose. But the Foundation did not deny that Tesla and Edison had been first choices.

Edison's fame and wealth were secure; he had little need of such an honor. But for Tesla it must have been one more cruel disappointment. And certainly it was not the kind of publicity he needed at this critical time.

20. FLYING STOVE

The teething troubles that beset the development of the new turbine were substantial. Elated with the initial success of his small turbine models, Tesla had designed a large double turbine to test with steam at the Waterside Station in New York. This was Edison country, peopled with engineers of the New York Edison Company, and predictably there were problems almost from the start.

Tesla's habit of arriving at the station sprucely attired at 5 P.M. and insisting that the workers stay for overtime caused no pleasure. There wasn't enough money to test the turbine properly, even on a straight schedule. The engineers, failing to understand it, reported it a misconception. And so on.

More important, there was a severe practical problem. At the tremendously high speeds at which the turbine operated, averaging 35,000 revolutions per minute, the centrifugal force was so great that it stretched the metal in the rotating disks. It was to be many years before metallurgy would produce the superior metals required.

He finally persuaded the Allis-Chalmers Manufacturing Company in Milwaukee to build three turbines, but again he was most undiplomatic with both engineering staff and management, and communicated his dissatisfaction to the board of directors. He walked out on tests after learning of a negative report by the engineers, claiming they would not build it as he wished. They said he refused to supply enough information.

When the manager of Westinghouse's railway and lighting division wrote asking for details on the turbine, Tesla replied confidently that it was superior to anything in the competition in terms of extreme lightness and high performance. Indeed, he said, he was planning to use it in a box-like flivver airplane.

"You should not be at all surprised," he wrote, "if some day you see me fly from New York to Colorado Springs in a contrivance which will resemble a gas stove and weigh as much."[1] (The plane would weigh only eight hundred pounds and could if necessary enter and depart through a window.)

This vision, however captivating, failed to bring Westinghouse orders. Accordingly, in his efforts to continue development of the turbine, he took the, for him, unusual step of working directly *for* two companies—the Pyle National Company and the E. G. Budd Manufacturing Company.

With the turbine he had invented a valvular conduit that enabled it to be used with combustible fuel. This unique conduit, with no moving parts, has recently been used in fluid logic elements, in which context it is referred to as a fluid diode.[2] Tesla's 1916 patent of his valvular conduit,* which closely followed Fleming's vacuum diode, is one of the cornerstones of the modern science of fluidics. But once again, he would manage to profit very little by his discovery.

Today the Tesla turbine is at last beginning to get some of the attention it has long deserved. One of the country's leading research experts on it is Professor of Engineering Warren Rice of Arizona State University who, however, has confined his work to the fluid mechanics of the flow processes that occur between the disks.[3]

In 1972 Walter Baumgartner built an experimental model of the Tesla turbine engine that ran on compressed air aided by steam injection, and produced some 30 horsepower, at 18,000 rpm.

In the 1980s the unique turbine is under active development for vehicular and power-plant use by SunWind, Ltd., of Sebastopol, California. SunWind, Ltd., plans to use a modified version of the Tesla turbine, burning hydrogen as the optimal fuel, in a three-wheel car called the Rainbow. The turbine will also burn propane, vegehol, and gasoline.

SunWind president Mark Goldis states that researcher Peter Myers

---

* Patent No. 1,329,559, valvular conduit; 1,061,142, fluid propulsion; 1,061,206 turbine. Also filed in the period 1909–1916: 1,113,716, fountain; 1,209,359, speed indicator; 1,266,175, lightning protector; 1,274,816, speed indicator; 1,314,718, ship's log; 1,365,547, flow meter; 1,402,025, frequency meter.

has built an experimental model of the turbine which verifies that it performs as Tesla predicted it would. He is now working on incorporating a proprietary combustion chamber to bring it up to the needs of contemporary designs and taking into account modern metallurgy.

"We are convinced that the improved Myers vortex turbine, based on Tesla's invention, will work better than any now in use, and that it will operate at 60 percent efficiency," Goldis says. Efficiency of most other turbines is about 40 percent. He believes that most earlier experimenters failed in their efforts to build Tesla's turbine because they did not understand laminar flow as opposed to turbulent flow. The turbine, says Goldis, is inexpensive and easily machined.[4]

Another California firm, General Enertech of San Diego, is building and selling the Tesla turbine as a pump. This too has been improved and modernized.

Alas, future vindication does not pay current bills. Tesla was having a hard struggle to meet the costs of day-to-day operation and to keep his credit for entertaining at Delmonico's. It came as a trifling blow, a social anticlimax, when for the second time he was dropped for nonpayment of dues by the Players' Club. With both Mark Twain and Stanford White gone, his pleasure in going to the old haunt had diminished.

Still, his name continued to appear regularly in the press, headlines never ceasing to proclaim the originality of his imagination. His ideas had news value even when substance was lacking. "Tesla's Tidal Wave to Make War Impossible," declared the *English Mechanic & World of Science*, disclosing his idea for the use of explosives to create destructive ocean waves upon demand. Little more was ever heard of this brainchild.

In a letter to the *Times* under the heading "Nicola [sic] Tesla Objects," the new vulnerable and touchy Tesla issued a generalized complaint to the effect that he thought he should receive credit for his own inventions. Shortly afterward the sour-grapes attitude that his friends had marked with sadness was betrayed again on the editorial page of the *Times* in parallel columns—one a letter from Tesla, the other a story about the hero of the hour, Orville Wright.

Wright was being interviewed in a flat meadow near Washington, D.C., as he prepared to take up his plane, which he had now flown many times, on a test flight. This threatened to be a special occasion, however, for word had been brought to Wright that President Teddy Roosevelt was standing by in the White House, hoping to be invited to accompany him as the nation's first Flying President.

Wright may be forgiven a certain nervousness at the thought of having for a passenger the toothy president, bundled from head to toe in high boots, leggings, helmet, goggles, and white silk scarf. It was a proper dilemma, as the *Times* report hinted. The flyer had wanted no part of such responsibility, knowing how genuinely risky the test could be. Yet it also seemed risky to have to say no.

A crowd of thousands had gathered on the crude flight strip awaiting the flyer's decision. Wright had spent as much time as was decently possible tinkering with the motor. At last the pioneer aeronaut raised his wind gauge aloft and studied it. The crowd held its collective breath. A slight zephyr fanned their brows. Wright lowered the gauge, shaking his head. "We cannot attempt a flight," he said gravely.

One column over, Tesla made clear his contempt for such a state of aeronautics. All his life he had been working on designs and engines for advanced high-speed planes, but thus far he had filed no patents. But he did not think much of what the competition had been doing and was at his most irritatingly superior:

"Place any of the later aeroplanes beside that of Langley, their prototype," he wrote, "and you will not find as much as one decided improvement. There are the same old propellers, the same old inclined planes, rudders, and vanes—not a single notable difference. . . . Half a dozen aeronauts have been in turn hailed as conquerors and kings of the air. It would have been much more appropriate to greet John D. Rockefeller as such. But for the abundant supply of high-grade fuel we would still have to wait for an engine capable of supporting not only itself but several times its own weight against gravity."[5]

The Langley plane, he said, was doomed if it encountered a downdraft, and the helicopter was in this respect much preferable, although objectionable for other reasons.

The really successful heavier-than-air craft would be based on radically new principles, he predicted, and would soon materialize. "[W]hen it does it will give an impetus to manufacture and commerce such as was never witnessed before, provided only that Governments do not resort to methods of the Spanish Inquisition, which have only proved so disastrous to the wireless art, the ideal means for making man absolute master of the air."[6]

Although such letters throbbed with the injuries done to him and only created more resentment toward him, his prophecy was, as usual, accurate. Honored at a dinner at the Waldorf with Rear Admiral Charles Sigsbee, he described the "aerial warships" that were coming and once more predicted a wireless telephone that would encircle the globe.

The patents on his brilliantly designed flivver airplane or flying stove—in today's technical literature the descendants of this craft (not to be confused with simple helicopters) are called vertical takeoff and landing aircraft (VTOL)—would not be filed until 1921 and 1927 and finally granted in 1928.[7] This is believed to be the only invention patented by Tesla of which, probably for lack of developmental capital, he built no prototype. The year the patents were issued the inventor would have been seventy-two years of age.*

The tiny plane, which he thought should sell for less than $1,000, rose straight into the air with its helicopter-type lifting propeller. The pilot touched a tilting device that pitched the craft forward, placing the propeller in front, airplane-style. The pilot's seat swivelled to remain upright while he moved the wings into a horizontal position. Tesla's light but powerful turbine was to thrust the plane forward at great speed. It could land by reversing the process—on a space the size of a garage roof, a living room, or the deck of a small boat.

Tesla's vertical-takeoff concept languished until nearly a decade after his death. Then, in the early 1950s, both Convair and Lockheed tested

---

* Patent No. 1,655,114, Apparatus for Aerial Transportation.

vehicles that, although vastly more sophisticated in engineering, adhered faithfully to the Teslian fundamentals. The more successful of these craft, the Convair XFY-1 "Pogo," was a 14,000 pound single-seat Navy fighter powered by a 5850 hp Allison T-40 turboprop engine. At rest, it sat on its tail, nose pointed skyward. In action, it took off vertically, then rotated 90 degrees to horizontal flight, in which it had a designed top speed in excess of 600 miles-per-hour at 15,000 feet.

Although tests of the "Pogo" were generally successful, the Navy decided not to put the plane into production. The Allison engine, Navy evaluators felt, was insufficiently powerful; the design of the pilot's pivoting seat was inadequate to accommodate the radical changes of attitude required, and the tricky, essentially blind, landings were just too dangerous.

But the potential military and commercial advantages of a full-scale aircraft that could take off and land without benefit of extensive runways were too great to be ignored. Following the intriguing tests of the Convair and Lockheed machines, the international aerospace industry entered into a full-scale pursuit of the ideal VTOL design. Numerous ideas were tried, but by the beginning of the 1980s the favored design was of an aircraft which did not itself change attitude on landing and takeoff, but whose engines were modified so that the direction of thrust could be rotated through 90 degrees. Two of the modern world's leading operational fighter planes—the Anglo-American British Aerospace "Harrier" and the Russian Yakovlev Yak-36—employ this principle.

Plainly Tesla's flivver-cum-flying stove was a far cry from today's sophisticated, massively powerful VTOL's. Indeed, conceived as it was decades before the advent of the jet engine, the flying stove could hardly have been otherwise. But as the Convair and Lockheed experiments of the 1950s suggest, the Teslian concept was an almost inevitable first step in true VTOL research. That Tesla should have hit upon this idea at a time when the enterprise of aviation was in its infancy is astonishing enough, but if we can credit the Yugoslav magazine *Review*, Tesla's VTOL concept may even have *anticipated* the advent of powered flight.

According to this generally respected publication, there is information in the Tesla papers in Belgrade indicating that Tesla's first VTOL drawings, along with plans for rocket motors, were destroyed in the laboratory fire of 1895![8]

The Nikola Tesla Museum in Belgrade contains, in addition to drawings of the aircraft, plans for an "aeromobile," a jet-propelled automobile with four wheels, apparently designed for flying or for terra firma. His papers, according to museum officials, include "calculations for horsepower, fuel and other aspects, all of which lost their true value when Tesla passed away." In addition, they report that he left sketches of interplanetary ships. This information, however, has not been made available to western scholars.

In more down-to-Earth moments, Tesla designed specially mounted lightning rods and air conditioning systems, and wrote proposals for manufacturers demonstrating that his turbine could be operated on the waste gases from steel mills and factories. He never saw smoke escaping from a stack when he was not offended by the waste of uncombusted fuel that used up finite resources.

While his imagination continued to soar with the future, the circumstances of his present became drearier by the day. A rare quarrel over money occurred between the inventor and Scherff, but was soon forgiven. Scherff wrote that creditors were "hounding me hard," and that the illness of his wife had put him in debt. He hoped Tesla would make some payment on his loans.

The inventor loftily responded, "Please do not give way to bitterness. You know that the experiences you have had were unusual and that while they have not benefitted you materially to a great extent, they have been the means of developing the good that is in you...."[9] When Scherff proved more insistent than usual, he sent a small amount of money and again took a superior stance in the matter: "I am sorry to note that you are losing your equanimity and poise.... You must pull yourself together and banish the evil spirits...."

To further bolster the morale of his former employee and loyal friend, he reported that the development of his steam and gas turbines and of a blower had been almost completed and that they held revolutionary promise. "I am now at work," he wrote, "on new designs of automobile, locomotive, and lathe in which these new inventions of mine are embodied and which cannot help but prove a colossal success. The only trouble is where and when to get the cash, but it cannot last very long before my money will come in a torrent, and then you can call on me for anything you like."[10]

On another occasion the much-tried Scherff pointedly wrote to say he was glad to hear a Tesla therapeutic device would soon be on the market because he himself could use one. Rather late in life, he bought a modest home at Westchester, Connecticut, and meeting his mortgage payments became a recurring subject with respect to Tesla's outstanding notes.

Although the "torrent" of money never came, Tesla did manage to find occasional major investors. Thus the Tesla Ozone Company was incorporated in 1910, with a capital of $400,000, to develop a process with several commercial uses, among them refrigeration. Later, the Tesla Propulsion Company was capitalized at Albany, New York, for $1 million by the inventor with Joseph Hoadley and Walter H. Knight, its purpose being to build turbines for ships and for the Alabama Consolidated Coal & Iron Company.

To add to his other problems, Tesla had trouble with his former employee, Fritz Lowenstein, in this period. Ever since the days of his secret research in Colorado, the inventor had worried about Lowenstein's loyalty. He was reassured when the German engineer returned to work for him at Wardenclyffe, but within a few years this relationship was terminated for financial reasons. Lowenstein went on to become a successful inventor of radio devices.

In 1916 he was called as a key witness for the defendants in the case of the *Marconi Wireless Telegraph Company of America* v. *Kilbourne and Clark*, having agreed to testify that in his opinion the Tesla radio patents held sway over the Marconi patents. At the last moment, however,

Lowenstein switched sides and testified for Marconi. Many questions were raised about his veracity, and charges were made but nothing was ever proved. As a result of this, however, he incurred Tesla's lasting enmity. It appeared that in the period 1910 to 1915 Tesla had lent substantial sums of money to the German radio engineer. Three years later Tesla brought suit against him, but did not go to trial.[11]

Anne Morgan, now famous in her own right, reappeared in his life tangentially after her father's death. Tesla had written to her of his deep admiration for the elder Morgan, which had outlasted his disappointments over money: "All the world knew him as a genius of rare powers, but to me he appears as one of the colossal figures of the ages . . . which mark epochs in the evolution of human thought and endeavor. . . ."[12]

Like Tesla's turbine, Anne had become a powerhouse, her life crowded with humanitarian activities in education, children's affairs, women's working conditions, and immigrant welfare—not to mention fashion and the servant problems of the wealthy. Fresh from the pleasures of touring Europe, she would turn up in Women's Night Court in Manhattan to befriend a wayward girl. An early Frances Perkins–without-portfolio, she traveled about America, speaking before women's clubs in behalf of her causes, which now included a vacation savings fund for working women. She conferred with judges about the problems of homeless, exploited young women, which were real and appalling, and sometimes she ranged as far as Topeka, Kansas, where Governor W. R. Stubbs once admiringly described her as an "insurgent."

Although she had all but forgotten her youthful infatuation with Tesla, they kept in touch. "I have hopes to see you this winter," she wrote, "and am indeed sorry that a whole year has passed since we last met. Have the months done much for you in your work, and do you now, at last, feel you are advancing . . . ?"[13]

Tesla, glad for the opportunity to renew their friendship, bragged a little: "The progress since our last pleasant meeting was steady and most gratifying. My ideas come in an uninterrupted stream as ever before. I see them grow and develop and am achieving happiness and, in a degree, success in the worldly sense." He praised her own "noble work" and sent warm regards to Mrs. Morgan.[14]

The Triangle Factory Fire of March 25, 1911, in which 145 shirtwaist workers, most of them young immigrant women, leaped to their death from a New York high-rise sweatshop, caused an outpouring of anger that led to more rapid unionization and ultimately to widespread reform of working conditions. Many additional workers had been injured in this fire, which resulted from a flagrant disregard for safety regulations. From this pivotal catastrophe much that Anne had worked for as a young woman was materializing.

She was seen marching with strikers and had become a writer of formidable letters in behalf of her causes. In her tailored suits she was what journalists described as "full-figured," a chainsmoker, fast talker, and much sought after as a fund raiser. It was said that her energetic presence "charged the atmosphere like an electrical disturbance."

One biographer has speculated that Anne's androgenic characteristics and Tesla's putative asexuality might have formed the basis of their friendship. Undoubtedly, however, money and social position formed a stronger magnetism.

In view of the many pleas for capital that Tesla made to her father and brother over the years, there is a certain turnabout humor in the fact that Anne did not scruple to tap him for her causes and that she appealed shrewdly to his snobbishness in doing so. In a long letter to Tesla while a fund raiser for the Women's Department of the National Civic Federation, she grouped her subjects under such headings as "Almshouses" and "Citizenship," reporting indignantly that the proponents of compulsory State Old Age Pensions had declared the almshouse "a relic of barbarism, a useless evil." No flaming liberal despite her support of the downtrodden, she believed that the government must save and improve the almshouses. In sly conclusion she asked, "Will you be one of thirty to contribute $100 towards the amount still needed this year? . . ."[15] There is no record that Tesla responded. He often had trouble paying his hotel rent.

In his desk lay another letter unanswered from Katharine Johnson: "Sometimes I hope you will make me tell you what I know about thought transference," she wrote. "One would need to feel herself *en rapport* to speak of such things. I have had such a wonderful experience the past three years, so much of it is already dim, that I sometimes fear it will all pass away with me and you of all persons ought to know something of it for you could not fail to have a scientific interest in it. I call it thought transference for want of a better word. Perhaps it is not all that. I have often wished and meant to speak to you of this but when I am with you I never say the things I had intended to say, I seem to be capable of only one thing. Do come tomorrow, Saturday." [16]

21. RADAR

The humiliating news of Tesla's financial distress following his loss of Wardenclyffe was further advertised in March 1916 when he was summoned to court in New York for failure to pay $935 to the city in personal taxes.[1] Scherff had lain awake nights worrying about his former employer and his taxes, and now it had happened. Every local newspaper carried the story. The misfortune seemed unjustly cruel, coming at a time when Edison had just been appointed to an important defense research post in Washington, while Marconi, Westinghouse, General Electric, and thousands of lesser firms were thriving on the profits from Tesla's patents.

He was now forced to confess in court that he had lived for years on credit at the Waldorf-Astoria, that he was penniless and swamped with debts. The land on which Wardenclyffe stood was taken from him and sold to a New York attorney, and it was even reported that the inventor might go to jail for contempt in connection with his tax debts.

Yet somehow in this time of turmoil and heartsickness he polished and published the basic principles of what would be known—almost three decades later—as radar.

German U-boats were sinking almost a million tons of Allied shipping a month when America entered World War I in April 1917, and the search for a way to detect submarines was of the highest priority. But there was as yet no such urgency about finding a means of predicting air attacks, although long-range German planes and Zeppelins had begun to raid central France and England with some regularity. Although it was predictable that aerial bombardment would eventually become horribly destructive, it was not yet so; and anyway, the air war was still thought to be romantic and dashing, bringing out a latent propensity for heroism even among its victims.

When German planes dropped the first bombs on Paris, Parisians stood in the open streets to watch. When London was attacked from the air, Londoners trampled primroses and hedgerows racing to the scene. An airship brought down in flames was described by a newspaper as "beyond doubt the greatest free show that London has ever enjoyed."

Even the bombing victims showed few signs of stress, said *The Lancet*, so unique and stimulating was the experience. In fact the English welcomed the chance to show what the reporter described as "a fundamentally important factor, that of race, [which] is seen *par excellence* in the response of the crowd to stimuli of the character that we have become familiar with since the outbreak of the War. . . ." War made the English feel more English.

In the circumstances, it is not surprising that when Tesla first began to speculate about military applications of radar, it was with respect to locating ships and submarines rather than to detecting enemy bombers. Tesla had predicted the general concept of radar in his sweeping article for *Century* magazine of June 1900: "Stationary waves . . . mean something more than telegraphy without wires to any distance. . . . For instance, by their use we may produce at will, from a sending station, an electrical effect in any particular region of the globe; we may determine the relative position or course of a moving object, such as a vessel at sea, the distance traversed by the same, or its speed. . . ."

In *The Electrical Experimenter* of August 1917 he described the main features of modern military radar: "If we can shoot out a concentrated ray comprising a stream of minute electric charges vibrating electrically at tremendous frequency, say millions of cycles per second, and then intercept this ray, after it has been reflected by a submarine hull for example, and cause this intercepted ray to illuminate a fluorescent screen (similar to the X-ray method) on the same or another ship, then our problem of locating the hidden submarine will have been solved.

"This electric ray would necessarily have to have an oscillation wave length extremely short and here is where the great problem presents itself, i.e., to be able to develop a sufficiently short wave length and a large amount of power. . . .

"The exploring ray could be flashed out intermittently and thus it would be possible to hurl forth a very formidable beam of pulsating electric energy. . . ."

What he had described were the features of atmospheric pulsed radar that would finally be practically developed in a crash program only months prior to the beginning of World War II.* Tesla intended it to be used as underwater radar, however, which later proved impracticable because of the great attenuation of electromagnetic waves in water. Despite much recent research, no means have yet been found of propagating light, high-frequency radio beams, or radar through the ocean. But Tesla's extra-low-frequency (ELF) waves *will* penetrate the seas and may serve a different purpose (see chapter 30), that of communication.†

Even if Tesla's radar could not be used to locate submerged objects, it was curious that no one could then imagine any other use for it. At least as far as the Navy was concerned, Edison may have had a hand in shunting radar aside. Now a white-haired elder statesman of invention, he had been named to direct the new Naval Consulting Board in Washington, with the primary job of finding a way of spotting U-boats. Tesla's idea, if even brought to Edison's attention, would almost certainly have been discounted as mere dream stuff.

In any event Edison had his hands full feuding with the Navy bureaucracy and cold-shouldering the "perfessers" who had begun clamoring for a piece of that new taste treat, the federal research pie. Edison's own ideas were repeatedly chopped down by the Navy brass while he suffered frustration. As it turned out, the negative

---

* A prototype of radar was officially credited to England's Robert A. Watson-Watt in 1935. But the history of modern microwave radar dates from 1940 when the multicavity magnetron became available. (*Encyclopaedia Britannica*).
† A blue-green laser communication system from Lawrence Livermore Laboratory shows similar promise.

ramifications of his appointment were to prove more important to history than anything positive he was able to do in the post.

At the time that Edison went to Washington, rumpled but rich, and Tesla remained in New York, poor but dapper, both men were aware that a gap as broad as the Hudson River was widening between them and the new generation of atomic physicists. The latter could talk of nothing but Einstein. The new people were specialists, although the splintering of minds was still in the infancy of its glory. They joined the American Physics Society and believed little that failed to appear in their journal.

Michael Pupin had gone to the trouble of carving out a section for engineers in the National Academy of Sciences, which previously had refused to admit even Edison. The line between practical men (engineers) and theoreticians (physicists) caused artificial distinctions to be drawn that were handicapping the war effort. Those who were inventors, scientists, *and* engineers, like Pupin and Tesla, or chemists *and* inventors like Edison, were almost by definition passé.

The new physics boiled with debates over waves versus particles and about Einstein's special theory of relativity, which Tesla—with strong cosmic theories of his own—rejected outright. When Einstein's general theory of relativity was published in 1916, even its creator had been unable to accept fully the dynamic universe that it implied. So disturbed was Einstein by this that he built into his calculations a "fudge factor" that preserved the possibility that the universe might after all prove to be stable and unchanging. To Tesla this was just added proof that the relativists didn't know what they were talking about. He himself was working on a theory of the universe to be disclosed in good time, and he had long ago propounded (but not published) his own dynamic theory of gravity.

He believed and had often stated, that atomic power would be 1. a dud, or 2. impossibly dangerous to control. In this he had illustrious company. Einstein too had grave doubts about it. As late as 1928 Dr. Millikan said, "There is no likelihood man can ever tap the power of the atom. The glib supposition of utilizing atomic energy when our coal has run out is a completely unscientific Utopian dream. . . ."[2] And even in 1933 England's Lord Rutherford could say, "The energy produced by the breaking down of the atom is a poor kind of thing. Anyone who expects a source of power from transformation of these atoms is talking moonshine."[3]

Perhaps it rankled Tesla to hear one of the "new physics" quips being attributed to Professor Sir William Bragg, co-winner of the 1915 Nobel Prize that for a time he had thought to be his. God runs electromagnetics on Monday, Wednesday, and Friday by the wave theory, said Bragg; and the devil runs it by quantum theory on Tuesday, Thursday, and Saturday.

Tesla's thoughts in later life were tending more and more toward a unifying physical theory. He believed that all matter came from a primary substance, the luminiferous ether, which filled all space, and he stoutly maintained that cosmic rays and radio waves sometimes moved more swiftly than light.

The younger scientists, most of whom were affiliated with universities, were just beginning to perceive what a garden of earthly delights government-sponsored research could be. Oddly enough it was to be Edison, creator of the modern industrial research laboratory, who threw a spanner into their dreams.

His first utterance as head of the Naval Consulting Board was that he did not think "scientific research would be necessary to any great extent." After all, he said, the Navy already had access to a vast "ocean of facts" in the Bureau of Standards. What the Navy needed was practical men to produce the technology, not theoreticians. And although the board was to have included civilian experts, he made it clear that he wanted no physicists—although a mathematician or two might be of some use.

The scientifically ambitious naval officers were as disconcerted as the university scientists. What about submarine detectors? they wanted to know. Wouldn't this take intensive research?

Edison, unperturbed, said he thought the whole idea of a Navy research laboratory too exotic. But if the Navy insisted upon it, he believed it should know how he handled things in his laboratories: "We have no system; we have no rules, but we have a big scrap heap." And inventors who circled around the scrap heap long enough usually came up with inventions. He did not mention that his own staff routinely referred to his laboratory as "the dungyard."

This was enough to drive the university scientists to action. They formulated a scheme that began with bypassing the Navy and aiming straight for the top. Through the National Academy of Sciences they appealed to President Wilson. The academy, they argued compellingly, could provide "an arsenal of science" for the country.

Soon the National Research Council, the ancestor of all subsequent research agencies, the fountainhead of science grants, was quietly formed. The NRC was to include leading scientists and engineers from universities, industry, and the government, with the goal of encouraging both basic and applied research. The second unerring move of the professors—which also set a precedent—was to establish headquarters in Washington, D.C., only blocks from the White House and Congressional purse strings.

The value of a National Research Council to corporate America was obvious. The group at once drew support from business and industry. A powerful pattern for the future had been delineated, the incestuous triumvirate of government, industry, and academe that would shape every aspect of American life in the twentieth century. And, ironically, it got started mainly as a tactic for circumventing "the old curmudgeon."

The government at once assigned the NRC the job and funds for discovering a way to detect marauding U-boats—the same job Edison's board was already working on. An Allied mission was also formed, with French and American scientists both racing to invent submarine-listening devices.

Tesla, his description of the future radar officially ignored, could not be bothered with such petty concerns as listening devices. Guided missiles and doomsday machines were more in his line. He gave *The New*

York Times a provocative peek at his latest patent applications for a new device "like the thunderbolts of Thor," capable, he said, of destroying whole fleets of enemy warships, not to mention armies.[4] "Dr. Tesla insists there is nothing sensational about it," reported the *Times*, "that it is but the fruition of many years of work and study."

He described the device as a missile that would zoom through the air at 300 miles per *second*, an unmanned craft with neither engine nor wings, sent by electricity to drop explosives at any point on the globe. Tesla said he had already constructed a wireless transmitter sufficiently powerful to perform this feat, but that it was not yet the time to disclose the details of his guided missile.

Nor had he given up on his scheme for creating fleets of robot warships. Just the year before he had urged the government to "install along both of our ocean coasts, upon proper strategic and elevated points, numerous wireless controlling plants under the command of competent officers and that to each should be assigned a number of submarine, surface, and aerial craft. From the shore stations these vessels . . . could be perfectly controlled at any distance at which they remained visible through powerful telescopes. . . . If we were properly equipped with such devices of defense it is inconceivable that any battleship or other vessel of an enemy ever could get within the zone of action of these automatic craft. . . ."

Washington could not have been less interested. All ears, it seemed, were cupped to the rather primitive listening devices being produced by NRC scientists, multiple-tube arrangements with electrical amplifiers designed for the hulls of submarine-detecting craft. These worked to a certain degree. Much later, when sonar was developed, the basic principles would be closer to Tesla's unsung concept of radar, for it would detect the presence of subs, mines, and the like by means of inaudible, high-frequency vibrations reflected back to the sending device from the targets.

By the war's end Edison, like Tesla, was thoroughly disillusioned with what he deemed the blindness and lack of creativity of the defense bureaucracy. Of the many projects that he had proposed, not one had been approved by the Navy Department.

Long after World War I, and fifteen years after Tesla's description of radar had been published, both American and French teams were diligently working to develop such a system according to his principles. Lawrence H. Hyland and Leo Young, two young scientists in the Naval Research Laboratory, rediscovered the potential application of high-frequency beams of short pulses of energy, this time with both aircraft and surface shipping in mind.

The military development of radar in America was to be impeded even further by interservice secrecy, but in time both the Army and Navy developed crude long-wave radar sets (one to two meters as opposed to microwaves). Meanwhile, in 1934, a French team under Dr. Emil Girardeau built and installed radar on both ships and land stations, using "precisely apparatuses conceived according to the principles stated by Tesla," says the Frenchman. "On the subject of Tesla's recommendation concerning the very great strength of the impulses," he added, "one must also recognize how right he was"; but the technology had been unavailable and "the most difficult thing was to succeed in enormously increasing the strength."[5]

In America the first seagoing radar tests were made in 1937 on the USS *Leary*, an old destroyer of the Atlantic fleet and their success led to development of the model XAF. A later model was in service on nineteen ships by 1941 and made an excellent wartime record.

Simultaneously an English team was struggling with this problem, for by now Hitler threatened England with invasion in World War II. The early pre-microwave radar installations used by the British Home Chain had very large antennae transmitting radio waves some 10 meters in length. Even so, these primitive sets were credited with winning air battles. Finally a sufficiently powerful magnetron was built which became the basis of all the generators established for modern radar starting with the 1940s.

German scientists also developed a form of radar. It was thus an international achievement inspired by the mind of Tesla, although the English scientist Robert A. Watson-Watt was officially credited with the invention in 1935.

The long race was won just in time to help save Britain from destruction by Nazi bombers in the Battle of Britain. Radar became the basic defensive tool of almost every country in the world. After the war it was eagerly employed by commercial airlines and shipping and would soon become essential to space exploration.

Dr. Girardeau says that at the time Tesla was formulating his principles, "he was prophesying or dreaming, since he had at his disposal no means of carrying them out, but one must add that if he was dreaming, at least he was dreaming correctly."[6]

At the time when his description of this invention appeared in print in 1917, Tesla was in Chicago. Broke but undefeated, he had again resolved to concentrate on developing his more practical inventions. Just before he left on this prosaic and arduous mission—painful for him, since it meant both dealing with engineers for a long period of time and being away from his friends—he was asked by one of his oldest admirers, B. A. Behrend, to accept what any other engineer in America would have deemed a high honor—the Edison Medal of the American Institute of Electrical Engineers.

It was as if Behrend had opened one of those overhead hotel fire extinguishers and it had rained down vitriol instead of water.

22. THE GUEST OF HONOR

B. A. Behrend was an engineer of great distinction and was himself in line to receive the prized Edison Medal. But he also felt keenly the injustices done to Tesla.

It was outrageous, he believed, that the man who had created the modern age of electric power, with all its blessings to people and industry, towns and cities around the world, should now be struggling to keep a hotel roof over his head. It was outrageous that he was being deprived of reward and honor for his invention of radio while others commercialized it; that he had received little credit for lighting inventions that were profiting others; that electrotherapeutics, adapted by more practical men from his high-frequency apparatus, was growing into a field of medical technology that seemed to benefit almost everyone but the inventor. And just the year before, Dr. Edwin Northrup had gone back for inspiration to the old ideas and circuits of Tesla to devise his first high-frequency furnace, a debt that he at least had graciously acknowledged. Behrend the engineer enumerated to himself only the more prosaic of Tesla's achievements and felt outraged.

He quickly found that persuading the AIEE to confer the Edison Medal upon Tesla was easy compared to getting the inventor to accept it. He did not want the Edison Medal. He would *not* receive it.

"Let us forget the whole matter, Mr. Behrend," he said. "I appreciate your good will and your friendship but I desire you to return to the committee and request it to make another selection. . . . It is nearly thirty years since I announced my rotating magnetic field and alternating-current system before the Institute. I do not need its honors and someone else may find it useful."[1]

The old wounds, reopened, bled bitterness. How indeed could the AIEE have been so remiss? More than three-quarters of the members of the Institute probably owed their own jobs to Tesla's inventions.

Since the hostility between Edison and Tesla was well known, it probably had been assumed that he might feel a certain distaste for the medal's name; but Behrend, knowing that the inventor both needed and deserved such acclaim at this period, insisted.

That brought down the rain of acid.

"You propose," said Tesla, "to honor me with a medal which I could pin upon my coat and strut for a vain hour before the members and guests of your Institute. You would bestow an outward semblance of honoring me but you would decorate my body and continue to let starve, for failure to supply recognition, my mind and its creative products which have supplied the foundation upon which the major portion of your Institute exists."[2]

It was rare for Tesla to reveal personal feelings toward Edison but now he pulled no punches. "And when you would go through the vacuous pantomime of honoring Tesla you would not be honoring Tesla but Edison who has previously shared unearned glory from every previous recipient of this medal."

Behrend, however, refused to let the matter rest there. After several visits to Tesla's office, he persuaded him to accept the honor.

Tesla passed the Engineers' Club almost daily, but no longer went inside. The building stood, as it still does, directly across from Bryant Park, the rectangle of sooty grass and listless trees behind the public library where he went each day to feed his pigeons. Many engineers observed the strange tall figure, less magnificently dressed than in his prime yet still erect and proud, as he entered the park to be greeted by swirls of birds. Pigeons even then were considered socially unmeritorious. Their hunger seemed to touch only people who were, like them, in need. Pigeons appealed to quirky, lonely, unreliable, usually poor, and eccentric persons. Important engineers did not hang about in city parks feeding dirty birds.

Journalists too had noticed Tesla on his avian missionary work.

Going home after midnight a reporter might find him standing in the
darkness, lost in thought, with a bird or two taking food from his hands
or lips, even though it was well-known that birds were blind at night and
preferred to be in their roosts. At such times Tesla was apt to make it
clear to the reporters that he did not care to talk with them. Later two of
them would find out why.

Another journalist told of meeting him wandering about in Grand
Central Station. When asked if he had a train to catch, he replied, "No,
this is where I do my thinking."

On the night of the Edison Medal presentation ceremony, a banquet
was held in the Engineers' Club. Afterward the members and guests
were to reconvene across the alley in the United Engineering Societies
building on 39th Street for speeches.

It was a splendid white-tie affair. The guest of honor was impecca-
ble, the radiance of his personality shining forth as forcefully as in his
youth. All eyes followed his tall, charismatic presence. Yet somehow
between the banquet hall and the nearby auditorium, he vanished.

How such a flagpole figure managed to disappear, Behrend could
not for the life of him understand. The committee was in a dither, and a
search was begun for the guest of honor. Waiters peered into rest rooms.
Behrend, thinking Tesla might have become ill, rushed into the street to
take a taxi to Tesla's hotel, the St. Regis. But following an impulse, he
found his steps turning instead toward Bryant Park.

Making his way through the gathering dusk, Behrend reached the
entrance to the park, only to find it blocked by a group of strollers
watching something in the shadows. Behrend edged his way in, and there
stood Tesla festooned from head to toe in pigeons. They perched upon
his head, pecked feed from his hands, and covered his arms, while a living, gurgling carpet of birds swarmed over his black evening pumps. The
inventor spotted Behrend and cautiously raised a finger to his lips, disen-
gaging feathered friends in the process.

Finally, while Behrend stood anxiously by, Tesla dusted feathers from
his finery and consented to be led back into the hall to receive his tribute.

Behrend's formal testimonial to his old friend was eloquent and sincere:

"Were we to seize and eliminate from our industrial world the results of Mr. Tesla's work," he reminded his colleagues, "the wheels of industry would cease to turn, our electric cars and trains would stop, our towns would be dark, our mills would be dead and idle. Yes, so far reaching is his work that it has become the warp and woof of industry. . . . His name marks an epoch in the advance of electrical science. From that work has sprung a revolution. . . ."

He closed by paraphrasing Pope's lines on Newton:

"Nature and Nature's laws lay hid in night;
God said, Let Tesla be, and all was light."[3]

The guest of honor found himself warming to the assemblage. He was after all human, and it was right and proper that these words should be spoken. He was pleased when W. W. Rice, Jr., president of the AIEE, reminded the audience of the scientific progress that had flowed from Tesla's research in oscillating currents.

"From his work followed the great work of Roentgen, who discovered the Roentgen rays," said Rice, "and all that work which has been carried on throughout the world in following years by J. J. Thomson and others, which has really led to the conception of modern physics. His work . . . antedated that of Marconi and formed the basis of wireless telegraphy . . . and so on throughout all branches of science and engineering we find . . . important evidence of what Tesla has contributed. . . ."[4]

The guest of honor rose at last with applause in his ears and found within him the power to speak graciously of Thomas Edison. He recalled his first meeting with "this wonderful man, who had had no theoretical training at all, no advantages, who did all himself, getting great results by virtue of his industry and application. . . ."[5]

Moving on he spoke rather longer than the engineers had expected, describing his childhood and later life, telling humorous anecdotes, and revealingly explaining "why I have preferred my work to the attainment of worldly rewards. . . ." Tesla said that he was deeply religious, although not in the orthodox meaning of the word, and gave himself "to the constant enjoyment of believing that the greatest mysteries of our being are still to be fathomed and that, all the evidence of the senses and the teachings of exact and dry sciences to the contrary notwithstanding, death itself may not be the termination of the wonderful metamorphoses we witness.

"I have managed to maintain an undisturbed peace of mind, to make myself proof against adversity, and to achieve contentment and happiness to a point of extracting some satisfaction even from the darker side of life, the trials and tribulations of existence. I have fame and untold wealth, more than this, and yet—how many articles have been written in which I was declared to be an impractical unsuccessful man, and how many poor, struggling writers, have called me a visionary. Such is the folly and shortsightedness of the world! . . ."[6]

Some years later Dragislav Petković, visiting from Yugoslavia, would walk with the inventor to Bryant Park on his daily mission of mercy and hear a revealing comment.

"Mr. Tesla looked up at the [library] windows, which are fenced with the iron bars, that some pigeons did not fall down somewhere and got freezed," he recalled. "In one corner he spotted one which was halfway frozen. He told me to stay here and watch that the cat does not come to get him while he look up for others. While I was watching, I tried to reach the pigeon, but could not do it because the bars were so close to one another. When Mr. Tesla returned, he quickly bended and pull him out.

" 'All things from childhood are still dear to me,' " he told Petković, as he began to pat the almost frozen pigeon, assuring it that it would recover.

"Then," said Petković, "he took the package from my hand and started throwing the food all around in front of the library. When he distributed the food he told me: 'These are my sincere friends.' "[7]

• • •

With the business of the Edison Medal over, Tesla entrained for Chicago
and devoted the remainder of the year to efforts to develop a variety of
inventions—not only in America but in Canada and Mexico. Thus he
hoped to make up for his wartime losses of European royalties.⁸ The previous year a trial balance of the Nikola Tesla Company had shown capital stock worth $500,000, laboratory expenses of $45,000 and patent
expenses of $18,938. Scherff, preparing his tax returns on a weekend,
reminded the inventor that the government could now fine him $10,000
for failure to file. If there was a net profit that year, Scherff failed to mention it in his letter.⁹

From his headquarters at the Blackstone Hotel, Tesla went to work,
offering not merely his inventions but himself as a consultant. A major
offering was his bladeless fluid turbo-generator for lighting systems,
small, simple, and unusually efficient, as the prospectus stated, an apparatus of "overwhelming superiority."

He had licensed his automobile speedometer to the Waltham Watch
Company, only to see auto manufacturing halted by the war. Nevertheless during 1917 he had an income of $17,000 in speedometer and
locomotive-headlight royalties.

He struggled over a report for the National Advisory Committee
for Aeronautics, hoping to supply the government with a small aircraft
motor one-fifth the weight of the Liberty motor then used. An exchange
of correspondence with NACA (the predecessor of NASA) failed to
result in a contract.

To Scherff, when he could spare a few moments from arduous days,
he scribbled that his research on a new wireless transmitter that would
render messages absolutely secret "will secure for the U.S. an overwhelming advantage in the great conflict as well as in peace. . . ."¹⁰ At the same
time he was promoting the Tesla Nitrates Company, the Tesla Electro
Therapeutic Company, and the Tesla Propulsion Company. The former,
based on an electrical process for making fertilizer from nitrates (nitric
acid) captured from the air (which he had alluded to in the *Century* magazine article of 1900) proved to be economically impractical.

Determined to escape from debt, he also maintained at long distance a laboratory for turbine work at Bridgeport, Connecticut. There he had contracted as well with the American & British Manufacturing Company to erect two wireless stations. Unfortunately these Wardenclyffe-type enterprises failed for lack of adequate capital.

No one could any longer claim that Tesla was not commercializing. He made money on some of these enterprises—not spectacular amounts but enough to begin paying off his debts to Scherff and to keep a staff.

To Johnson, now harried by creditors, he wrote: "Write your splendid poetry in serenity. I will do away with all your worries. Your talent cannot be turned into money, thanks to the lack of discernment of the people of this country, but mine is one that can be turned into carloads of gold. I am doing this now." [11]

Johnson became ill. He wrote to remind Tesla of an old debt of $2,000, and the inventor at once sent a check for $500. Two weeks later Robert again wrote that he needed funds, this time for taxes, and Tesla sent another $500. Before the year was out Robert sent an SOS saying that he had only $19.41 in his bank account, with outstanding debts of $1,500. Once again Tesla reached for his checkbook. [12]

In his desk in New York lay a letter, some years old, from Katharine Johnson, one of the last that was kept, or perhaps written, by her to her "ever silent friend." She had gone to Maine without her children or husband for part of the summer.

"I came here a month ago, quite alone," she wrote, "to this hotel full, but empty for me, since it is a strange world. Here, I am as detached as if nothing belonged to me but memory. At times I am filled with sadness and long for that which is not—just as intensely as I did when a young girl and I listened to the waves of the sea, which is still unknown, and still beating about me. And you? What are you doing? I wish I could have news of you my ever dear and ever silent friend, be it good or bad. But if you will not send me a line, then send me a thought and it will be received by a finely attuned instrument.

"I do not know why I am so sad, but I feel as if everything in life had slipped from me. Perhaps I am too much alone and only need companionship. I think I would be happier if I knew something about you. You, who are unconscious of everything but your work and who have no human needs. This is not what I want to say and so I am Faithfully yours, KJ."[13]

She added a postscript: "Do you remember the gold dollar that passed between you and Robert? I am wearing it this summer as a talisman for all of us."

Money? Good fortune? A return to the happiness and excitement of earlier days? Would it be a talisman for the trio that had shared so much?

23. PIGEONS

People speak of decades as if they form natural endings, when in fact they seldom end anything cleanly. Human survivors are dragged into new slices of time with which they feel no harmony and in which they are often exposed to rasping change. So it was for Tesla in the Roaring Twenties.

The twenties brought the hypocrisy of Prohibition. A dignified man could no longer walk into his favorite bar and order a drink, but was instead forced to resort to illegal rotgut, bathtub gin, or worse. Speakeasies and gangsters flourished. Flaming Youth and bead-twirling flappers danced the nights away to the Charleston; the stock market alternately soared and dived, while speculators made and lost fortunes. James J. Walker, the Whoopee Mayor of New York, was one of those attuned to the times. Nikola Tesla, Victorian in manner and appearance, was not. He was, if anything, more estranged than ever from the world about him.

Hobson, who had been a Congressman and was soon to be honored with the Congressional Medal (carrying the rank of rear admiral) for his courage during the Spanish-American War, had lost his recent bid for the U.S. Senate. But he had not lost—to Tesla's intense regret—his campaign against drink and had been instrumental in obtaining passage of the Eighteenth Amendment. To Tesla Prohibition constituted an intolerable bureaucratic invasion of personal liberty. He freely expressed his opinion that it would shorten lives, including his own. He no longer could foresee living until the age of 140. Without the divine ambrosia in modest but regular amounts, who would care to?

Yet when the Hobson family returned to Manhattan to live, Tesla was well enough pleased that he and the sometime hero could be close again. Hobson took up the reins of other worthy campaigns, including leadership of an international commission on narcotics, but he always found time for his old friend. He began the habit of hunting Tesla up in his hotel once a month to attend a movie matinee. It was a curiously frivolous diversion for such a distinguished pair. They would emerge from stale darkness into the glare and clangor of a Times Square afternoon and move off to a favorite park bench. There they would talk of world politics and science, or reminisce about old times.

Now in his mid-sixties, Tesla was almost always hard up. At times strange illnesses troubled him. The businesses he had worked so hard to build up in Chicago were dwindling away. Wardenclyffe was no more than a sad memory, yet he never ceased to strive for the development of his world wireless system. In 1920 he again approached Westinghouse executives with a wireless proposal. Their rejection brought from him a tart reminder that, at the time of obtaining rights to his alternating-current system, the directors had promised him that "nothing will be turned down that you may put before the Westinghouse Company." He had relied upon their assurance, he said, "knowing that men of that stature usually feel a sense of obligation to the pioneer who lays the foundations to their successful business...."[1]

He found the firm's attitude doubly frustrating because they were in fact now entering the wireless field, and Tesla had heard that they planned to put up a broadcasting system. "In the first place I was astonished and keenly disappointed," he wrote, "that the matter should have been put before your engineers.... I would never submit anything to them except complete plans, thoroughly worked out in every detail...." Westinghouse officials responded by offering him a temporary consulting job.

The following year Westinghouse inadvertently insulted him by writing that they had begun operation of a Radio-phone Broadcasting System at Newark, New Jersey, presenting news broadcasts, concerts, and crop and market reports; and inviting him as a guest to speak to their "invisible audience."[2] Haughtily he reminded them that he had long worked to develop a broadcasting system to encompass the globe: "I prefer to wait until my project is completed before addressing an invisible audience and beg you to excuse me."[3]

At the same time, however, he again offered Westinghouse the designs of his "commercially superior turbine," which he assured them would save the firm millions of dollars. But he warned that there could be no strings. He could produce the turbines at once but would not consent to agree to "any experimenting whatever."[4] The response was tiresomely familiar. Board chairman Guy E. Tripp wrote that they could not enter such an agreement because their engineers were negative on the subject, "and of course *we must be guided by the opinion* of our Engineers."[5]

Two special friends entered Tesla's life in this period, a sculptor and a writer, whose respective talents would help to preserve his name and achievements from the obscurity that could befall even a famous person who had neither heirs nor a corporate identity to prod the public's memory. The nineteen-year-old science writer, Kenneth M. Swezey, arrived on the scene to join the ranks of the inventor's permanent coterie; and the Yugoslav sculptor, Ivan Meštrović, middle-aged and already famous in Europe, came to New York to introduce his work to America.

Tesla and the sculptor cherished common memories of their childhoods in the mountains of Yugoslavia. Both were poets at heart. They met often in New York, talking about anything and everything. Both worked late into the night and had a similar problem. Meštrović was forced to wrangle his hunks of marble from one hotel to another for lack of a studio; Tesla, to his great sadness, could no longer afford a laboratory. So they took long walks together, discussed Balkan affairs, their work, and shared their pleasure in reciting Serbian poetry. Along the way, Meštrović was introduced to the daily routine of feeding the pigeons of Manhattan.

Long after the sculptor had returned to Split, Tesla at the urging of Robert Johnson wrote and asked him to do a bust of himself. He could not go to Europe however, and Meštrović was unable to return to America. Nevertheless, the latter wrote back, saying that he remembered the inventor so well that, if Tesla would send a photograph, he would undertake the job.[6] Tesla replied that he had no money; Meštrović answered that none was needed. Good as his word, he sculpted and cast in bronze a powerful and sensitive likeness (now to be seen at the Tesla Museum in Belgrade) that transcended the miles, the years, and mere realism to capture the brooding essence of genius.*

As for young Swezey, on meeting the inventor for the first time in 1929, he was surprised to discover (as he wrote) "a tall skinny man of upright posture" who might go about for hours in a daze of concentration, but who also had a side intensely human and "almost painfully sensitive with fellow-feeling for everything that lives."[7]

Swezey himself, residing in a bleak apartment in Brooklyn, had few close ties to family or friends. He became both a journalistic champion of the scientist and a devoted admirer. The old man and the younger were often together. Although Tesla worked hard while others slept, he also knew how to refresh himself with long rambles through the city. Swezey often joined him on these nocturnal excursions.

He too was introduced to the pigeons. One evening as they were walking down Broadway, with Tesla discoursing intensely on his system for sending electrical power wirelessly to the ends of the Earth, the inventor suddenly lowered his voice. "However, what I am anxious about at this moment," he said, "is a little sick bird I left up in my room. It worries me more than all my wireless problems put together."

The pigeon, which he had picked up two days before in front of the library, had a crossed beak which had started a cancerous growth on its tongue so that it could not eat. Tesla had saved it from slow death and said that with patient treatment it would soon become strong and well.

But not all of the birds he saved could be fitted into his hotel room, where the servants already complained of dirt. "In a large cage in a bird shop," wrote Swezey, "are several dozen more pigeons. . . . Some had wing diseases, others broken legs. At least one was cured of gangrene,

* A duplicate was also cast in bronze on Meštrović's order, which may be seen at the Technical Museum in Vienna. It was unveiled June 29, 1952, by Tesla's nephew, Sava Kosanović.

which the bird specialist pronounced incurable. If a pigeon is afflicted with something that Tesla has not the facilities to treat, it is put under the care of a competent physician."

He and Swezey, as they walked, talked of Einstein, diet, exercise, fashion, marriage. "Tesla's only marriage has been to his work and to the world," wrote the young man, "as was Newton's and Michelangelo's ... to a peculiar universality of thought. He believes, as Sir Francis Bacon did, that the most enduring works of achievement have come from childless men...."[8]

The inventor confided to his young companion that mental anguish, fire, commercial opposition, and other trials had merely fanned his productiveness and that he still felt he could rise highest in the face of great resistance. He also said that he had earned in his lifetime over $2 million. Yet, for him to have earned this sum he probably would have to have received the legendary $1 million for his alternating-current patents from Westinghouse.*

Because so many strange interpretations have been made of Tesla's devotion to pigeons, the following letter from Tesla to Pola Fotić, the young daughter of Konstantin Fotić, Yugoslavian ambassador to the United States, is cited for its simple portrayal of love for the creatures of his childhood. Entitled "A Story of Youth Told by Age," he describes the winter isolation of the house where he was born, and of his special friend, "the magnificent Mačak, the finest of all cats in the world."[9]

It was in connection with Mačak that his first intimation of electricity came to him one snowy evening when he was three years of age. "People walking in the snow left a luminous trail behind them," he wrote, "and a snowball thrown against an obstacle gave a flare of light like a loaf of sugar hit with a knife...." Even at that early age his vision was hyperreceptive to light. Footprints in the snow were not in muted shades of blue, purple, or black as they might seem to others.

"I felt impelled to stroke Mačak's back. What I saw was a miracle

---

* Much later, after the inventor's death, Swezey made a careful effort to verify this story by examining the Westinghouse archives. He could find nothing to support it.

which made me speechless. . . . Mačak's back was a sheet of light, and my hand produced a shower of crackling sparks loud enough to be heard all over the place."

His father told him this was caused by electricity. His mother said to stop playing with the cat lest he start a fire. But the child was thinking abstractly.

"Is nature a gigantic cat? If so who strokes its back? It can only be God, I concluded."

Later, as darkness filled the room, Mačak shook his paws as though he were walking on wet ground, and the boy distinctly saw the furry body surrounded by a halo like the aura of saints. Day after day he asked himself what electricity could be, and found no answer. At the time of writing this letter, eighty years had gone by, and Tesla said that he still had no answer.

In contrast to the cat's delightful company was the family gander— "a monstrous ugly brute, with a neck of an ostrich, mouth of a crocodile and a pair of cunning eyes radiating intelligence and understanding like the human." In old age Tesla claimed to have a scar inflicted by the monstrous bird. But the other creatures on the farm he loved.

"I liked to feed our pigeons, chickens, and other fowl, take one or the other under my arm and hug and pet it." And even the vicious gander, when it brought its flock home at night after "sporting like swans" in a meadow brook, "was a joy and inspiration to me." Now, in New York, as he withdrew more and more from a frenzied age and from people with whom he felt little harmony, his fondness for pigeons took on a strange intensity.

He became alarmingly ill in his office on 40th Street one day in 1921 and, as usual, refused to see a doctor. When it became apparent that he might be unable to return to his apartment at the St. Regis Hotel, he whispered to his secretary to telephone the hotel, speak with the house-keeper on the fourteenth floor, and tell her to feed the pigeon in his room—"the white pigeon with touches of gray in her wings."[10] He insisted that the secretary repeat this urgent message after him. The housekeeper was to continue feeding the pigeon each day until further notice. She would find plenty of feed in the room.

Whenever in the past the inventor had been unable to visit Bryant Park with the feed, he had hired a Western Union messenger to take care of the errand for him. The white pigeon, it was apparent, was special to him. From his attitude, his secretaries thought he might be delirious.

He recovered, and the matter was forgotten—until another day, when he telephoned his secretary to say the pigeon was very ill and that he could not leave the hotel. Miss Skerritt recalled that he spent several days at home. When the pigeon had recovered, he resumed his usual routine of working, walking, thinking, and feeding the birds.

About a year later, however, he arrived at his office looking shaken and distraught. In his arm he carried a tiny bundle. He summoned Julius Czito, who lived in the suburbs, and asked if he would bury the dead pigeon on his property, where the grave could be properly cared for. But scarcely had the machinist returned home on this curious mission than he received a phone call from Tesla, who had changed his mind.

"Bring her back, please," he said, "I have made other arrangements." How he finally disposed of her, his staff never knew.

Three years later Tesla was completely broke and his bill at the St. Regis Hotel had gone unpaid for a long time. One afternoon a deputy sheriff arrived at his office and began seizing his furnishings to satisfy a judgment against them. Tesla managed to persuade the officer to grant him an extension. When he had gone, there remained the matter of his secretaries, who had received no salaries in more than two weeks. All that was left in his Mother Hubbard's cupboard of a safe was the gold Edison Medal, which he now removed. It was worth about one hundred dollars, he said to the embarrassed young women. He would have it cut in two and give half to each.

Dorothy Skerritt and Muriel Arbus declined in one voice. They offered instead to share with him the small sums of money in their own purses.[11] When Tesla was able to pay them a few weeks later, he placed an additional two weeks' salary in each envelope. Yet on the day when he had offered to divide up the Edison Medal, there had in fact been a little money in the office—$5 in petty cash. But this he claimed at once for his pigeons, saying he was out of bird seed. He had asked one of his secretaries to go out and buy a fresh supply.

With the help of Czito, to whom he also owed a substantial amount of money, he then moved all his office belongings into a new office building. The next blow fell shortly afterward when he was asked to vacate the St. Regis Hotel, in part because of his pigeon friends. At one point Tesla had put some of the birds into a hamper and sent them home with patient George Scherff, thinking that a spell in Connecticut might do them good. But alas, so fond were they of their old friend and of their risky old haunts that they were back on his window ledge in time for dinner.

Sadly he packed up his possessions of decades and moved to the Hotel Pennsylvania. The pigeons followed. After another few years, he and they would be forced to move on to the Hotel Governor Clinton. Nikola and his birds were to spend the final decade of his life in the Hotel New Yorker.

The strange tale of the white pigeon was told by the inventor to O'Neill and William L. Laurence, science writer for *The New York Times*, one day while the three sat in the Hotel New Yorker lobby. John O'Neill, a member of a psychic society, saw mystic symbolism in Tesla's white pigeon. He and other psychics who have written about the inventor preferred to speak of the pigeon as a dove. Although pigeons are technically rock doves, only the most meticulous birdwatchers ever call them that and Tesla never called his pigeon anything but a pigeon. But what he told the two journalists in the hotel lobby, says his early biographer, was the dove love-story of his life.

"I have been feeding pigeons, thousands of them, for years," he said. "Thousands of them, for who can tell—.

"But there was one pigeon, a beautiful bird, pure white with light gray tips on its wings; that one was different. It was a female. I would know that pigeon anywhere.

"No matter where I was that pigeon would find me; when I wanted her I had only to wish and call her and she would come flying to me. She understood me and I understood her.

"I loved that pigeon.

"Yes, I loved her as a man loves a woman, and she loved me. When she was ill I knew, and understood; she came to my room and I stayed beside her for days. I nursed her back to health. That pigeon was the joy of my life. If she needed me, nothing else mattered. As long as I had her, there was a purpose in my life.

"Then one night as I was lying in my bed in the dark, solving problems, as usual, she flew in through the open window and stood on my desk. I knew she wanted me; she wanted to tell me something important so I got up and went to her.

"As I looked at her I knew she wanted to tell me—she was dying. And then, as I got her message, there came a light from her eyes—powerful beams of light."

Tesla paused and then, as if in response to an unasked question from the science writers, continued.

"Yes, it was a real light, a powerful, dazzling, blinding light, a light more intense than I had ever produced by the most powerful lamps in my laboratory.

"When that pigeon died, something went out of my life. Up to that time I knew with a certainty that I would complete my work, no matter how ambitious my program, but when that something went out of my life I knew my life's work was finished.

"Yes, I have fed pigeons for years; I continue to feed them, thousands of them, for after all, who can tell—."

The writers left him in silence and walked several blocks along Seventh Avenue without speaking.

O'Neill later concluded: "It is out of phenomena such as Tesla experienced when the dove flew out of the midnight darkness and into the blackness of his room and flooded it with blinding light, and the revelation that came to him out of the dazzling sun in the park at Budapest, that the mysteries of religion are built." Had Tesla not rigorously suppressed his mystical inheritance, he wrote, "he would have understood the symbolism of the Dove." [12]

Dr. Jule Eisenbud, in an article for the *Journal of the American Society for Psychical Research*, has examined the bird symbolism in the inventor's life in conjunction with his neuroses and his childhood relationship with his mother, to the extent that the latter is known. The bird is an age-old universal symbol of the mother and her nourishing breast, says the psychologist. And it was significant that Tesla believed he could command his beautiful white pigeon to appear, wherever he was, *with only his wish.* "The meaning of this fantasy," he asserts, "can be arrived at only when viewed in conjunction with the strong evidence from other biographical data that the unconscious need for, and for control of, the 'disappearing' mother had dominated Tesla throughout his life, accounting not only for many of his clinically peculiar habits, and much that was out of the ordinary in his relationship to people and things, but even for the private mythology in terms of which he seems unconsciously to have conceived the powerful all-pervading force he devoted his life to capturing and harnessing." [13]

Nothing in Tesla's writings indicates to the lay person that he felt deprived by a "disappearing" mother. But Dr. Eisenbud sees in his life many signs of an emotionally and physically deprived infantile nursing period. Tesla consciously idealized his mother, insists Eisenbud, yet he managed to stay clear of her, "and for most of his life was given to unfulfilled premonitions (all but the last unfulfilled, that is) of her death, her ultimate disappearance. This kind of ambivalence, the sort of thing seen frequently in persons who are known clinically as obsessional neurotics, which Tesla definitely was, marked all his relationships to and attitudes toward mother symbols and mother substitutes."

Thus, says Eisenbud, he could not tolerate smooth round surfaces, and pearls on a woman made him physically sick. He speaks of an obsessional patient of his own who, on his mother's testimony, had gone into a deathlike depression when abruptly taken off the smooth round breast at the age of two weeks and in later life could not stand even the word *sphere.*

Dr. Eisenbud believes the inventor's attitude toward money was also indicative of a deep-lying fantasy of having virtual control of this universal mother symbol at the source:

"He gave away millions in gestures of great, if sometimes bizarre generosity, and was often broke as a result. He was, however, apparently dominated by the comforting belief that fundamentally he was not dependent on fate or other people for his sustenance, and that money itself, a trivial and incidental aspect of the tedious mechanics of living, he could make in sufficient amounts whenever he needed it. . . . The most extraordinary facet of Tesla's never-ending game of control of the mother, however, was played out with food itself, where, unhappily, the negative side of his ambivalent attitude toward this most direct of all mother substitutes finally won out. . . ."[14]

Hence, he says, the elaborate ceremony Tesla made of dining, arriving in evening clothes at the appointed hour, to be shown to his special table, the head waiter becoming an expensive mother surrogate "the symbolic control of whom is not infrequently striven for by those in the chips."

He remarks on the fact that one of Tesla's favorite dishes was squab: "In a beautiful clinical example of biting the breast that didn't feed him (the other side of the coin of his compulsive feeding of pigeons) he would . . . eat only the meat on either side of the breastbone."

As the wheel of his life came full circle, says Eisenbud, Tesla was reduced to living mostly on warm milk. Then it was that his beautiful white pigeon "gave forth her last dazzling, blinding beam of light—a symbol equated with the stream of milk from the breasts. . . ."[15] Tesla's lifetime of compensation and *ersatz* collapsed. Something went out of his life, and he knew that his work was finished.

Behavioral theorists will argue with such Freudian/Jungian conclusions, however, tending to believe that specific traumatic incidents in Tesla's childhood, leading to emotional repression accounted for his obsessional neuroses.

Unfortunately a lack of conclusive data makes it impossible to do more than speculate.

24. TRANSITIONS

Katharine Johnson fell ill. Tesla showed his concern by prescribing for her a special diet, but the deeper illness from which she suffered, the sense that in the midst of life everything worthwhile had somehow slipped from her, deprived her of the will to recover. She lay in the house at 327 Lexington Avenue with the blinds drawn, remembering the parties, the celebrities, the gossip and reflected glory, the street crowded with arriving and departing hansom cabs and autos, the wonderful banquets presided over by Tesla at the Waldorf-Astoria, the thrill of his galvanic presence at her table, and how hard they had all worked to entrap wealthy patrons for him. She remembered the scintillating gatherings at his laboratory, the demonstrations, the excitement of his triumphal tours abroad. Her entire being seemed to have dissolved into a blur of memories. The life lived had not been hers; she did not know whose it had been. Her life had been a reflection only, of the risks and acts and triumphs of others. Now she felt a stranger to herself, stripped equally of hope and anger. She felt deluded, cheated, and infinitely weary.

During the time when she languished Tesla was brashly inspired to think about writing one of his more curious prophecies—on the future of women. It was a subject on which he gyrated and fussed and yet seemingly could not leave alone. The year before she was stricken, he had given an interview to a Detroit *Free Press* reporter on the "problem" of women.[1] With the glibness of any other male, he bemoaned their descent from the pedestals so thoughtfully built by men for their entrapment. He had worshipped women all his life, he said—out of special deference, from afar. But now that they were matching their minds against men's,

venturing into open competition with God's naturally appointed, was
not "civilization itself in jeopardy"? The answer was a question that presumably went unasked by most Sunday supplement readers of the 1920s:
Whose civilization?

Now, with Katharine's illness preying on his thoughts, he turned the
matter over relentlessly in his mind and finally gave another interview,
this time to *Collier's*.[2] The article was threateningly entitled, "When
Woman Is Boss," and described a new sex order in which the female
would emerge as intellectually superior. On the one hand he appeared to
be all for it, but on the other, filled with trepidation. Had he understood
the real waste of Katharine's life? Whatever his motivation, he ended up
ambivalently prophesying men and women into human beehives in a
disturbingly mechanistic view of the Utopian "rational" society.

It was clear to any trained observer, he said, that a new attitude
toward sexual equality had come over the world, receiving an abrupt
stimulus just before the First World War. Naturally he could not foresee that in the wake of the Second World War women would again
backslide and relinquish much social and economic gain in a compulsion to procreate.

Few feminists would have quarreled with the first part of Tesla's
premise: "The struggle of the human female toward sex equality will end
in a new sex order, with the female as superior. The modern woman, who
anticipates in merely superficial phenomena the advancement of her sex,
is but a surface symptom of something deeper and more potent ferment-
ing in the bosom of the race.

"It is not in the shallow physical imitation of men that women will
assert first their equality and later their superiority, but in the awakening
of the intellect of women.

"Through countless generations, from the very beginning, the social
subservience of women resulted naturally in the partial atrophy or at
least the hereditary suspension of mental qualities which we now know
the female sex to be endowed with no less than men.

"But the female mind has demonstrated a capacity for all the

mental acquirements and achievements of men, and as generations
ensue that capacity will be expanded; the average woman will be as
well educated as the average man, and then better educated, for the
dormant faculties of her brain will be stimulated to an activity that
will be all the more intense and powerful because of centuries of
repose. Women will ignore precedent and startle civilization with
their progress."

But the ideal society that Tesla went on to describe, modeled on that
of the hive—with "desexualized armies of workers whose sole aim and
happiness in life is hard work"—could not have failed to chill his fellow
men and thinking women.

"The acquisition of new fields of endeavor by women, their gradual
usurpation of leadership," he said, "will dull and finally dissipate feminine sensibilities, will choke the maternal instinct, so that marriage and
motherhood may become abhorrent and human civilization draw closer
and closer to the perfect civilization of the bee. . . ."[3]

The perfect communal life of the bee was radical chic for the times,
promising "socialized cooperative life wherein all things, including the
young, are the property and concern of all."

But in the same freewheeling interview Tesla made uncannily far-
sighted technological predictions: "It is more than probable that the
household's daily newspaper will be printed 'wirelessly' in the home
during the night. The problem of parking automobiles and furnishing
separate roads for commercial and pleasure traffic will be solved. Belted
parking towers will arise in our large cities, and the roads will be multiplied through sheer necessity, or finally rendered unnecessary when civilization exchanges wheels for wings.

"The world's internal reservoirs of heat . . . will be tapped for
industrial purposes." Solar heat would partially supply the needs of
the home; wireless energy would supply the remainder; and small
vest-pocket instruments, "amazingly simple compared with our present telephone," would be used. "We shall be able to witness and hear
events—the inauguration of a President, the playing of a World Series

game, the havoc of an earthquake or the terror of a battle—just as though we were present."

Katharine died in 1925. Not forgetting Tesla even at death, she charged Robert to keep in close touch with him always.

Johnson and his daughter Agnes (the future Agnes Holden) tried thereafter to celebrate traditional family holidays. Tesla was always invited. They invited him on Katharine's birthday, Robert writing, "We will have music, the kind of occasion she would have desired. She cherished your friendship. She charged me not to lose sight of you. Without you it will not be her day."[4]

But soon Robert was again asking for financial help—to pay taxes and a bank loan. Tesla, scraping along on a few royalties and consulting fees, was able to lend small sums. Although he had been ill again, he sent a cheerful note with his check: "Do not let those small troubles worry you. Just a little longer and you will be able to indulge in flights on your Pegasus."[5]

Johnson thanked him and announced that he and Agnes were sailing for Europe for two months. On this trip he met a teenaged actress who was to gladden for a time his final years.[6]

In April of the following year Tesla sent Johnson an unsolicited $500 with a note: "Please do not let this remind you of vulgar creditors, but have a little celebration."[7] Johnson replied that he was having a wall erected at Kate's grave with half of it. He reported that the "lovely Marguerite [Churchill]" was keeping him young and that he was eager for the inventor to meet her.

Shortly afterward Johnson was hospitalized and from his bed wrote Tesla: "You must come and dine with Mrs. Churchill and Marguerite when I return." He raved about the young actress, whom he now hoped to accompany to Europe, "with her mother, of course."[8] They would visit the homes and haunts of Tennyson, Keats, Shakespeare, Wordsworth. Instead, however, he went back to Europe with Agnes the following year and again in 1928, on both occasions with the aid of checks from the hard-pressed Tesla.

Francis A. Fitzgerald, who had been a personal friend of Tesla's since the development of Niagara Falls and who was with the Niagara Power Commission at Buffalo, tried to assist the inventor with one of his most cherished scientific concepts in 1927. He interceded with the Canadian Power Commission to finance a project to transmit power without wires. The venture was not carried out, but it planted in the minds of some Canadians a seed that regenerates itself every few years down to the present writing in efforts to transmit hydroelectric power wirelessly and inexpensively through the Earth.

For years it had been rumored that Tesla had invented a powerful beam, a death ray, but he had been strangely uncommunicative on the subject. In early 1924 a flurry of news reports from Europe claimed that a death ray had been invented there—first by an Englishman, then by a German, and then a Russian. Almost at once an American scientist, Dr. T. F. Wall, applied for a patent on a death ray which he claimed would stop airplanes and cars. Then a newspaper in Colorado proudly retorted that Tesla had invented the first invisible death ray capable of stopping aircraft in flight while he had been experimenting there in 1899.⁹ The inventor was unusually noncommittal on the matter.

In 1929 when Scherff again filled out tax returns for the Nikola Tesla Company he told Tesla: "Unfortunately the Company had no tax to pay." In this respect he was at least in tune with the times, for now the Great Depression had begun.

Tesla wrote another cheerful note to raise the spirits of his old friend Johnson, while yet admitting to his own "little financial fainting spells." He said, "Of course I am not communicative with other friends. My prospects are better . . . another very fine and valuable new invention." If he were one of the new inventors who employed press agents, he said, "the whole world would be talking about it."¹⁰

In fact, however, his patent filings had at long last almost dwindled to a stop. He had filed a series of new patents in 1922 in fluid mechanics, which were not processed to completion. Thus they entered the common domain. One among them is believed to have particular significance. Filed March 22, 1922, it was entitled "Improvements in Methods of and Apparatus for the Production of High Vacua."* Years later, when both the United States and Russia entered the race to perfect modern death/disintegrator ray weapons, it would be one of his ideas studied with special interest.

This was the first group of patents that he had filed since 1916. But if anyone were to have taken this as evidence that Tesla's creative life was drawing to a close, he would have been much mistaken.

---

* The others: Method of and Apparatus for Compressing Elastic Fluids; Method of and Apparatus for the Thermodynamic Transformation of Energy; Improvements in Methods of and Apparatus for Balancing Rotating Machine Parts; Improvements in Methods of and Apparatus for Deriving Motive Power from Steam; Improvements in Methods of and Apparatus for Economic Transformation of the Energy of Steam by Turbines; Improvements in Methods of Generation of Power by Elastic Fluid Turbines; Improvements in Apparatus for the Generation of Power by Elastic Fluid Turbines.

25. THE BIRTHDAY PARTIES

Born at midnight, never sure which date to celebrate, Tesla usually had not observed his birthdays at all. They had simply slipped by, and as long as he felt well, their passage had gone unnoticed.

He took pride in the fact that his weight had not changed since college days. Legends were told about his catlike fitness. Walking down Fifth Avenue one icy winter day, he had lost his footing, hurled himself into a flying somersault, landed on his feet, and kept on walking. Bug-eyed pedestrians swore that they had never seen anything like it outside of a circus.

But in old age he began to make up for the missed birthdays. Each anniversary became the occasion for a celebration with reporters and photographers. At these parties, to the delight of his young friends, he announced fantastic inventions and indulged in prophesy to his heart's content. Only sober Mr. Kaempffert, with the dignity of the *Times* to uphold, found such sessions grating. How they hung on the guru's every word as he spun his visionary nonsense. And worst of all, how they pretended to understand![1]

A very special birthday party was arranged by Swezey for Tesla's seventy-fifth anniversary. This shy young science writer was a person of few words—one who knew him remembers that he spoke almost cryptically—yet he was extraordinarily gifted in his ability to make science understandable to lay audiences by translating abstractions into graphic images. He made party games of science and thought up puzzles and simple kitchen-table experiments that captivated children. He wrote a book, *After-Dinner Science*, that enjoyed a popular success, especially with the parents of school-aged children. He also wrote advanced articles for scientific magazines.

Tesla was a hero to him. Swezey was, of course, more capable than the average person of appreciating the inventor's importance in the perspective of the history of science, and like Behrend, he was troubled by the public's short-mindedness. He resolved to do something about it.

And so, for the inventor's seventy-fifth birthday party in 1931, he asked famous scientists and engineers the world over to send some thing, and a flood of congratulatory letters and tributes to Tesla poured in. Among the authors were several Nobel laureates who acknowledged, with respect and gratitude, his inspiration to their own careers.[2]

Robert Millikan wrote of attending a Tesla lecture at the age of twenty-five, one of the first demonstrations of the Tesla coil. "Since then," he wrote, "I have done no small fraction of my research work with the aid of the principles I learned that night so that it is not merely my congratulations that I am sending you but with them also my gratitude and my respect in overflowing measure."

Arthur H. Compton declared: "To men like yourself who have learned first hand the secrets of nature and who have shown us how her laws may be applied by solving our everyday problems, we of the younger generation owe a debt that cannot be paid. . . ."

All the past presidents of the American Institute of Electrical Engineers sent tributes along with many leaders in the burgeoning field of modern radio.

Lee De Forest wrote of his deep personal obligation to Tesla as scientist and inventor: "For no one so excited my youthful imagination, stimulated my inventive ambition or served as an outstanding example of brilliant achievement in the field I was eager to enter, as did yourself. . . . Not only for the physical achievement of your researches on high frequencies which laid the basic foundations of the great industry of radio transmission in which I have labored, but for the incessant inspiration of your early writings and your example, do I owe you an especial debt of gratitude."

Dr. Behrend spoke of "the world's usual ingratitude toward its bene-factors.

"To those of us who have lived through the anxious and fascinating period of development of alternating-current power transmission," he said, "there is not a scintilla of doubt that the name of Tesla is as great here as the name of Faraday is in the discovery of the phenomena under-lying all electrical work."

Einstein, who seemed unaware of Tesla's prodigious range of achievement, sent his felicitations but congratulated him only on his con-tributions to the field of high-frequency current.

Among Europeans who sent accolades were Dr. W. H. Bragg, co-winner of the controversial 1915 Nobel Prize in physics. From the Royal Society in London he wrote, alluding to the demonstrations made by Tesla in his lectures forty years earlier:

"I shall never forget the effect of your experiments which came first to dazzle and amaze us with their beauty and interest."

Count von Arco, the German radio pioneer who with Prof. Adolf Slaby had developed the Slaby-Arco system wrote: "If one reads your works today—at a time when radio . . . has attained such a world significance—particularly your patents (practically all of which belong to the past century), one is again astonished at how many of your suggestions, often under another's name, have later been realized. . . ."

Swezey, the catalyzer of this outpouring of tributes, added his own most eloquently. Tesla's genius, he said, had given startling impetus to the work of Roentgen and J. J. Thomson and those who followed them in the age of the electron. "Standing alone," said the science writer, "he plunged into the unknown. He was an arch conspirator against the estab-lished order of things."

If these encomia seem immoderate, they pale beside the comments of famed science editor and publisher Hugo Gernsback: "If you mean the man who really invented, in other words, *originated* and discov-ered—not merely *improved* what had already been invented by others— then without a shade of doubt Nikola Tesla is the world's greatest inventor, not only at present but in all history. . . . His basic as well as revolutionary discoveries, for sheer audacity, have no equal in the annals of the intellectual world."

Alerted by Swezey to the birthday tributes, newspapers and magazines all over the world carried articles on Tesla. *Time* magazine's cover story reported that its writers had some difficulty tracking the elusive inventor ("a tall . . . eagle-headed man") to his most recent aerie at the Hotel Governor Clinton. Interviewers regretted they could not see him as he used to be seen in his Colorado laboratory, wrote *Time*, "strolling or sitting like a calm Mephistopheles amid blazing, thundering cascades of sparks . . . ."[3]

What they found instead was a Tesla emaciated and almost ghostlike but still alert. His hair was slate gray, his overhanging eyebrows almost black. But the sparkle of his blue eyes and the shrillness of his voice indicated his psychic tension.[4]

When Swezey presented the inventor with the bound memorial volume, he found him surprised but scarcely overwhelmed. Although he merely said that he did not care for compliments from people who all his life had opposed him, the young science writer felt that secretly Tesla was pleased by the many tributes. Indeed, later when Swezey tried to borrow them briefly (copies were sent to the new Tesla Institute at Belgrade), the old man was most reluctant to part with them.

To interviewers Tesla had disclosed the ideas that currently preoccupied his thoughts. He was working on two things: one, conclusions that tended to disprove the Einstein Theory of General Relativity. His explanations, Tesla said, were less involved than Einstein's, and when he was ready to make a full announcement, it would be seen that he had proved his conclusions.

Secondly, he was working to develop a new source of power. "When I say a new *source*, I mean that I have turned for power to a source [to] which no previous scientist has turned, to the best of my knowledge. The conception, the idea when it first burst upon me was a tremendous shock."[5]

He said of this new *source* of power that it would throw light upon many puzzling phenomena of the cosmos. And in another enigmatic comment that puzzles Tesla scholars down to the present day, he said it might prove of great industrial value "particularly in creating a new and virtually unlimited market for steel."[6]

Questioned further, he would only say that such power would come from an entirely new and unsuspected source, that for all practical purposes it would be constant day and night and at all times of the year. The apparatus for manufacturing this energy and transforming it would be of ideal simplicity with both mechanical and electrical features.

Tesla said the preliminary cost might be thought too high, but this would be overcome, for the installation would be both permanent and indestructible. "Let me say," he emphasized, "that [it] has nothing to do with releasing so-called atomic energy. There is no such energy in the sense usually meant. With my currents, using pressures as high as 15 million volts, the highest ever used, I have split atoms but no energy was released...."

Pressed to reveal his new source of energy, he politely declined, but promised definitely to make a statement on the subject "in a few months, or a few years."

His eyes glowing beneath the black brows, he said that he had already conceived of a plan for transmitting energy in large amounts from one planet to another—absolutely regardless of distance.

"I think that nothing can be more important than interplanetary communication," he said. "It will certainly come some day and the certitude that there are other human beings in the universe, working, suffering, struggling, like ourselves, will produce a magic effect on mankind and will build the foundation of a universal brotherhood that will last as long as humanity itself."

When? He was unsure.

"I have been leading a secluded life, one of continuous, concentrated thought and deep meditation," he replied. "Naturally enough I have accumulated a great number of ideas. The question is whether my physical powers will be adequate to working them out and giving them to the world...."[7]

Also in the seventy-fifth year of his life, *Everyday Science & Mechanics* carried detailed designs of two of the inventor's more down-to-earth proposals—a plan for extracting electricity from seawater and another for a geothermal steam plant.[8]

The geothermal steam plant was designed to draw upon the almost inexhaustible heat of the deep earth, with water circulating to the bottom of a shaft, returning as steam to drive a turbine, and then returned to liquid form in a condenser, in an unending cycle. Such ideas were not original with Tesla, having been speculated upon for at least seventy-five years, but he was among the first to draw up detailed designs.

His seawater power plant would utilize heat energy derived from the temperature differential between layers of ocean water to operate great power plants. He even went so far as to design a vessel to be propelled by energy derived from this source.

But his research into the matter was at best preliminary. He still had to overcome the same problems that earlier pioneers had experienced—great technological difficulties and costs far out of proportion to the greatest possible returns; yet he continued to work and improve the design, substituting for pipes hung in submarine abysses a sloping tunnel lined with heat-insulating cement. His associates, he said, had made studies in the Gulf of Mexico and Cuban waters where temperature contrast would be adequate.

Tesla explored several variations—one that operated without storage batteries; one that operated without water pumps—but he was still unsatisfied with his seawater plants, finding their performance too small to be competitive with other sources. Undaunted, he continued to predict that the technical problems were soluble and that one day such plants would be major producers of power.

Tesla did not live to see such a plant built, except in his mind. But in the 1980s the federal government has authorized a crash program of research on Ocean Thermal Energy Conversion (OTEC) plants in the

Gulf of Mexico, the Caribbean, Hawaii, and wherever temperature contrasts are adequate. A small army of university scientists appear to be employed on these joint ventures of government and private utilities.

Professor Warren Rice of Arizona State University, an authority on Tesla's work in turbines and fluid mechanics, has analyzed his anticipation of OTEC ideas and of geothermal energy recovery and finds them "thermodynamically sound." But he adds that he, personally, is pessimistic about the economic feasibility and practicality of OTEC and of terrestrial energy recovery on a large scale. "I hope that I am wrong," he adds.[9]

In his old age Tesla was gratified to hear his invention of electrical oscillation devices for medical therapy receive high accolades. At the American Congress of Physical Therapy in New York on September 6, 1932, Dr. Gustave Kolischer of Mount Sinai Hospital and Michael Reese Hospital, Chicago, announced that high-frequency electrical currents were bringing about "highly beneficial results" in dealing with cancer, surpassing anything that could be accomplished with ordinary surgery.[10]

Modern cancer treatment has, of course, progressed even farther and the full medical implications of Tesla's techniques are still being explored. Most recently, in the 1980s, the American Association for the Advancement of Science announced promising research in the electromagnetic stimulation of cells for the regeneration of amputated limbs. And studies at various universities have also indicated that pulsed current is superior to direct current in the healing of fractures.

As is typical of so many of Tesla's inventions, scholars still do not know the whole range of their possible applications or, in some cases, even their full theoretical significance.

26. CORKS ON WATER

George Sylvester Viereck was a German immigrant, the child of an illegitimate offspring of the House of Hohenzollern. He came to America in his youth, stirred the avant-garde with his precociously brilliant poetry, and became a controversial figure in politics and journalism. Intellectuals considered him a genius. But as his interviews with the rising stars of fascism, Hitler and Mussolini, betrayed his strong partiality for dictators, the poet's reputation was damaged, much as Ezra Pound's would be a few years later. The issue came to a head during World War II when Viereck was imprisoned for disseminating pro-Nazi propaganda.

He and Tesla became friends between the wars, the inventor as usual being politically uncritical. They often corresponded and met socially in New York. Viereck wrote insightful articles about Tesla, and the two exchanged their own poetry. The German's tenuous claim to royalty and his literary talent may have appealed equally to Tesla, who addressed some of his most revealing correspondence to this new confidante.

The only sample extant of the inventor's own poetry, called "Fragments of Olympian Gossip," and written in his distinctive hand, was dedicated to Viereck, "my Friend and Incomparable Poet," on December 31, 1934. Tesla was then seventy-eight years of age. The poem begins, "While listening to my cosmic phone / I caught words from the Olympus blown," which gives a fair indication of its literary merit. It is a crotchety work, but not without humor and occasional nice turns of phrase.

On April 7 Tesla wrote to Viereck, urging him to stop taking the "poison" of opium tincture, lest it make his precious brain sluggish. It appears that Viereck also was seeking escape from financial anxieties, for Tesla added: "It is too bad that the greatest poet of America is no better situated than a struggling inventor. How about writing a little article on Spiritism and drawing on my experience as told in a letter to you? The spiritists are so crazy that they will claim I got the message all right but as a crass materialist I was prejudiced. . . ."[1]

He added a P.S. that his admiration for Viereck was so great that his handwriting had even begun to resemble the poet's.

In December he wrote a long, strange letter to Viereck going back over the death of his brother Daniel so long ago, and the death of his mother. He attempted to explain away his precognition and discussed his affliction with partial amnesia. The letter was written as if from different time frames, without transitions, and with confusing errors as to Daniel's age when he died and the date of his mother's death. It is almost as though Tesla were describing dreams rather than reality.

He spoke of periods of tortured concentration driving him to fear a blood clot or atrophy of the brain, and of how he struggled to "drive out of the mind the old images which are like corks on the water bobbing up after each submersion. But after days, weeks or months of desperate cerebration I finally succeed in filling my head chuckfull with the new subject, excluding everything else, and when I reach that state I am not far from the goal. My ideas are always rational because I am an exceptionally accurate instrument of reception, in other words, a seer. But be this true or not I am always mighty glad when I get through for there can be no doubt that such a surtax of the brain is fraught with great danger to life."[2]

Viereck's writings—not so much in his correspondence as in his published work—also give us an interesting impression of what Tesla might have been thinking about at this time. In a 1935 magazine article entitled "A Machine to End War," Viereck reported on what Tesla believed the world would be like in the years 2035 and 2100.

"Man in the large," said the inventor, "is a mass urged on by a force. Hence the general laws governing movement in the realm of mechanics are applicable to humanity."[3]

He saw three ways in which the energy determining human progress could be increased: first, the improvement of living conditions, health, eugenics, etc.; second, reduction of the intellectual forces that impede progress, such as ignorance, insanity, and religious fanaticism; third, the enchaining of such universal sources of energy as the sun, ocean, winds, and tides.

He believed his own mechanistic concept of life to be "one with the teachings of Buddha and the Sermon on the Mount." The universe was "simply a great machine which never came into being and will never end. The human being is no exception to the natural order. Man, like the universe, is a machine. Nothing enters our minds or determines our actions which is not directly or indirectly a response to stimuli beating upon our sense organs from without. Owing to the similarity of our construction and the sameness of our environment, we respond in like manner to similar stimuli, and from the concordance of our reactions, understanding is born. In the course of ages, mechanisms of infinite complexity are developed, but what we call 'soul' or 'spirit,' is nothing more than the sum of the functionings of the body. When the functioning ceases, the 'soul' or the 'spirit' ceases likewise."[4]

Tesla pointed out that he had expressed these views long before the behaviorists, led by Pavlov in Russia and Watson in the United States, and stated that such an apparently mechanistic view was not antagonistic to an ethical or religious conception of life. In fact, he believed that the essences of Buddhism and Christianity would comprise the religion of the human race in 2100.

Eugenics would then, he believed, be firmly established. In a harsher time survival of the fittest had weeded out "less desirable strains," Tesla reasoned. "Then man's new sense of pity began to interfere with the ruthless workings of nature," and the unfit were kept alive. "The only method compatible with our notions of civilization and the race is to prevent the breeding of the unfit by sterilization and the deliberate guidance of the mating instinct. Several European countries and a number of states of the American Union sterilize the criminal and the insane."

How much of this pitiless doctrine was the aging Tesla and how much pure Viereck, one cannot say. Whoever was responsible, he was only just getting into his stride. "This is not sufficient," according to Tesla. "The trend of opinion among eugenists is that we must make marriage more difficult. Certainly no one who is not a desirable parent should be permitted to produce progeny. A century from now it will no more occur to a normal person to mate with a person eugenically unfit than to marry a habitual criminal." By 2035, a Secretary of Hygiene or Physical Culture would be more important than a Secretary of War.

Sounding rather more like the real Tesla, the putative Tesla goes on to foresee a world in which water pollution would be unthinkable, in which the production of wheat products would be adequate to feed the starving millions of India and China, in which there would be systematic reforestation and the scientific management of natural resources, in which there would at last be an end to devastating droughts, forest fires, and floods. And of course, the long-distance wireless transmission of electricity from water power would end the need to burn other fuels.

In the twenty-first century civilized nations would spend the greater part of their budgets on education, the least on war. He had at one time believed that wars could be stopped by making them more destructive. "But I found that I was mistaken. I underestimated man's combative instinct, which it will take more than a century to breed out. . . . War can be stopped, not by making the strong weak but by making every nation, weak or strong, able to defend itself."[5]

Here he was referring to a "new discovery" that would "make any country, large or small, impregnable against armies, airplanes, and other means of attack." It would require a large plant, but once established, it would be possible to "destroy anything, men or machines, approaching within a radius of 200 miles. It will, so to speak, provide a wall of power offering an insuperable obstacle against any effective aggression."

He explicitly stated, however, that his invention was not a death ray. Rays tended to diffuse over distance. "My apparatus," he said, "projects particles which may be relatively large or of microscopic dimensions, enabling us to convey to a small area at a great distance trillions of times more energy than is possible with rays of any kind. Many thousands of horsepower can thus be transmitted by a stream thinner than a hair, so that nothing can resist. This wonderful feature will make it possible, among other things, to achieve undreamed-of results in television, for there will be almost no limit to the intensity of illumination, the size of the picture, or distance of projection."[6]

It was to be not radiation but a charged particle beam. Almost half a century later the two most powerful nations in the world would be racing to perfect such a weapon.

Tesla also predicted that ocean liners would be able to cross the Atlantic at great speed by means of "a high-tension current projected from power plants on shore to vessels at sea through the upper reaches of the atmosphere." In this connection he alluded to one of his earliest concepts: such currents, passing through the stratosphere, would light the sky and to a degree turn night into day. It was his idea to build such power plants at intermediate points, such as upon the Azores and Bermuda.

The deepening political turmoil in Europe in the mid-1930s did not spare Yugoslavia. The Serbian ruler, King Alexander, who had established a Yugoslavian dictatorship following a move toward separatism by Croatia, was assassinated at Marseille in 1934 by a Croat terrorist.

Tesla promptly wrote to *The New York Times* in defense of the "martyred" monarch. Seeking to minimize the historic differences separating Serbs and Croats, he described King Alexander as "a heroic figure of imposing stature, both the Washington and Lincoln of the Yugoslavs . . . a wise and patriotic leader who suffered martyrdom."[7] It was true enough that there had never been unification of the Slaves until Alexander forced it upon them, but it would take another strongman (Tito) to make it stick.

Alexander was succeeded by his son, the young King Peter II, under the regency of Prince Paul. Tesla accordingly transferred his loyalty to the boy king, who would grow up prematurely in a world aflame.

Meanwhile, Franklin Delano Roosevelt had been elected President of the United States. Proclaiming a New Deal and calling Congress into a special session (the famous "100 Days"), he achieved passage in a short space of time of more long-lasting social legislation than had ever before been undertaken. In doing so he fused the rage of political opponents and drew charges of wanting to "pack" the Supreme Court. Tesla was one of those who, having voted for Roosevelt, soon found his socialistic whirlwind alarming.

More than ever, the inventor seemed obsessed with his mysterious new defensive weapon. In a last poignant appeal for capital to J. P. Morgan, he wrote: "The flying machine has completely demoralized the world, so much so that in some cities, as London and Paris, people are in mortal fear from aerial bombing. The new means I have perfected affords absolute protection against this and other forms of attack. . . .

"These new discoveries which I have carried out experimentally on a limited scale, created a profound impression. One of the most pressing problems seems to be the protection of London and I am writing to some influential friends in England, hoping that my plans will be accepted without delay. The Russians are very anxious to render their borders safe against the Japanese invasion and I have made them a proposal which is being seriously considered.

"I have many admirers there," he continued, "especially on account of the introduction of my alternating current system. . . . Some years ago Lenin made me twice in succession very tempting offers to come to Russia but I could not tear myself from my . . . work."[8]

Tesla went on to say that words could not express how he ached for a laboratory again and for the opportunity of squaring his account with the senior Morgan's estate. "I am no longer a dreamer but a practical man of great experience gained in long and bitter trials. If I had now $25,000 to secure my property and make convincing demonstrations, I could acquire in a short time colossal wealth. Would you be willing to advance me this sum if I pledged to you these inventions?"

He closed with an attack upon Roosevelt's program, no doubt calculated to soften Morgan: "The 'New Deal' is a perpetual motion scheme which can never work but is given a semblance of operativeness by unceasing supply of the people's capital. Most of the measures attempted are a bid for votes and some are destructive to established industries and decidedly socialistic. The next step might be the distribution of wealth by excessive taxing if not conscription. . . ."[9]

Morgan, who had his own Depression problems, failed to rise to the bait. For a nonscientist it was virtually impossible to tell whether Tesla was talking sense or nonsense.

He had made a similar offer of his "particle beam" to Westinghouse that spring, to which Vice-President S. M. Kintner had replied that he had discussed with a research specialist "the general proposal of creating rays of the kind you mention." But the specialist had been skeptical, "so much so in fact that I hesitate to propose to Mr. Merrick your suggestion of a six months' advance payment to enable you to file patents."[10]

Although it is always tempting to cast Tesla in the role of prophet without honor, it is conceivable that the research specialist was correct about the "particle beam." Tesla was perfectly capable of going off half-cocked, as his forays into metallurgy (in part the result of his dissatisfaction with the metals available for use in his turbine) suggest.

He formulated a process for degasifying copper (removing the bubbles to produce a superior metal) and interested the American Smelting and Refining Company (ASARCO) in it. Dr. Albert J. Phillips, then superintendent of the central research department of ASARCO, recalls meetings with Tesla on the project. In the depths of the Depression he would arrive at the firm's laboratories in Perth Amboy from the Hotel McAlpin in New York where he then lived, in a spendid chauffeur-driven limousine. He usually wore a frock coat, gray striped trousers, gray spats, and carried a cane with a gold knob.

"Dr. Tesla was a fine distinguished gentleman whom I liked very much," Dr. Phillips told me. "He was probably the world's greatest electrical theorist of the time. However, he was not a metallurgist and failed to realize that there was a great deal known about metals that he did not know. His experiments in the field of copper metallurgy were poorly planned and completely unsuccessful. Nevertheless I learned a great deal from my association with him and recall fond memories of his idiosyncrasies." [11]

The inventor's theory was that gas bubbles dispersed in a liquid were under pressures much higher than those computed by accepted theories, and he believed that such pockets of air or nitrogen, if small enough, would have the same density as copper in the liquid form. He arrived at the plant with complete drawings of an apparatus he wished built to prove his theory.

"I immediately informed him," recalls Dr. Phillips, "that the apparatus he had so carefully designed would not melt copper and could not possibly subject liquid copper to bombardment under vacuum to remove the hypothetical gas bubbles from it. I also told Dr. Tesla that there was plenty of evidence to prove that these hypothetical gas bubbles could not exist in molten copper to any great extent."

The two discussed their differences in a friendly scientific manner, "but [Tesla] was not swayed from his beliefs by my objections. . . ." So they proceeded to build the apparatus exactly as Tesla had designed it. And the results were just as the research superintendent had predicted. At last, liquid copper that had been melted elsewhere was poured into the equipment, subjecting the stream of metal to high vacuum and bombardment against a "Lava" target before it issued from the bottom into a mold.

"We finally obtained several samples of copper through the machine," Phillips recalls, "which instead of being densified were quite gassy and were in no way different from copper that had not been subject to the Tesla treatment."

And then, since the budget was badly overrun, the experiments were ended. To the best of Dr. Phillips' recollection, ASARCO had initially approved $25,000 for the venture ("In 1933 that was a lot of money and hard to come by") and may have extended it later by a similar amount. [12]

A curious detail emerged from these recollections. Tesla showed Dr. Phillips "a photograph of a cancelled check for $1 million, if I remember correctly, that he had received from the Westinghouse Electric Company for one of his patents or inventions." Since there is no record anywhere else of this check, the mystery of the payment for his alternating-current patents remains unresolved.

With occasional consulting jobs Tesla managed to survive the Depression and even lent small sums to friends in greater need. In one especially tight spot he went to Westinghouse and, for old times' sake, was given a job that brought in $125 a month for a brief period. Another time he turned to Robert Johnson and received help in his "temporary financial fainting spell," the latter replying from Stockbridge, Massachusetts, "I have in the bank $178. So I send you herewith $100. I hope that will do. Heaven bless you!"

Some time later Johnson fell ill. In his new "old" hand, he wrote: "At 83 I have just published my book, 'Your Hall of Fame.' . . . I shall not live to see your bust placed there . . . but there it will be, never doubt, my great and good friend . . .

"My heart is still yours for of all the years of friendship every day is dear.

"I am told that I am on the mend but the recovery is a long time coming . . ."[13]

Mend he did, however temporarily, for he was soon issuing an invitation to Tesla with a flash of the old gaiety: "Our ladies will wear their prettiest gowns and the gentlemen will dress in your honor tomorrow, and I suggest that you run true to form and look beautiful in evening dress for the ladies! I want them to see you at your handsomest. . . .

"Yours ever with remembrance of the happy old times, Luka J. Filipov."[14]

Then it was Tesla's turn to be ill. He had grown gaunt and gray, seldom leaving his hotel, subsisting on milk and Nabisco crackers. In his suite enameled empty cracker canisters, all neatly numbered, were stacked in rows on shelves. He used them as storage for odds and ends, as Swezey noted on his frequent calls. The latter was alarmed by the deterioration in the inventor's condition.

Johnson wrote: "God bless you and help you dear Tesla and may you recover to normal and to this end, do let us come to you. Agnes will be of great use. You have only to telephone. Do this in memory of Mrs.

Johnson. . . ."[15] But he himself had suffered a relapse and realized that the end was near. "Neither of us can count on many years," he wrote. "You have few friends besides the Hobsons and us to look after you. Do let Agnes come to you. I cannot. Not to do this will be suicide, dear Nikola."[16] Soon, however, the inventor had mended.

The year 1937 was to be one of sad losses for Tesla. Hobson, his staunch friend of many years, died suddenly on March 16 at the age of sixty-six.

Robert Johnson died on October 14, following recurring illnesses.

Shortly afterward, on a cold midnight, Tesla left the Hotel New Yorker on his regular rounds to scatter feed for his pigeons. Only two blocks from the hotel he was struck by a taxi and hurled to the street. Refusing medical care, he asked to be returned to his hotel room.

Although in a state of shock, he telephoned a Postal Telegraph messenger, William Kerrigan, to call for the pigeon feed and finish his errand. For the next six months, Kerrigan went daily to feed the flocks at St. Patrick's Cathedral and Bryant Park.

It was discovered that Tesla had sustained three broken ribs and a wrenched back. Complications from pneumonia followed, and he lay bedridden until the spring. Although he recuperated, his health remained even more frail thereafter, and he was subject to periods of irrationality.[17]

From old friends at the Westinghouse Company came word that the Tesla Institute, which had been founded two years earlier at Belgrade, Yugoslavia, was seeking information about his early inventions. Tesla agreed to have his photo taken beside his original split-phase alternating-current motor for the research laboratory that was being equipped in his honor at the institute.[18]

An endowment had been underwritten for this purpose by the Yugoslav government and individual Slavs, and it would include an honorarium for Tesla of $7,200 per year. Thanks to his native countrymen, "the greatest inventive genius of all time" would at least not be destitute in his final years.

27. COSMIC COMMUNION

"One hears many strange things about him," said Agnes J. Holden, the daughter of Robert and Katharine Johnson. "It's not right to judge a man who has passed eighty by what he did in his eighties. I remember Tesla when he was thirty-five years old, young and gay, and full of fun."

But the inventor at eighty still enjoyed life and was in fact still formulating his far-reaching statement on the universe. Looking forward to his birthday parties, he prepared papers for them months ahead and planned stunning headlines for his friends of the press. Increasingly, the parties were occasions for refuting Einstein, defending Newton, and advancing the cosmic theories that Tesla himself had long mulled over.

The ten-page statement he issued on his eightieth birthday in 1936 was never published in its entirety. Both in it, and in letters to the *Times*, he waged a continuing debate with leading physicists as to the nature of cosmic rays.[1]

He alluded often to his own dynamic theory of gravity, which he said would explain "the motions of heavenly bodies under its influence so satisfactorily that it will put an end to idle speculations and false conceptions, as that of curved space." In his considerable writing on astrophysics and celestial mechanics, however, this theory of gravity was never elucidated.

The curvature of space, he stated, was entirely impossible since action and reaction are coexistent. A curve would be counteracted by straightening. Nor would any explanation of the universe be possible without recognizing the existence of ether and its indispensable function. The Einsteinian revolution notwithstanding, he remained convinced that there was "no energy in matter other than that received from the environment." And this, he held, applied rigorously to molecules and atoms as well as to the largest heavenly bodies.

In short, he was quite wrong.

For the occasion of his eightieth birthday he spoke of yet more inventions for interstellar communication and energy transmission.

"I am expecting to put before the Institute of France an accurate description of the devices with data and calculations and claim the Pierre Guzman Prize of 100,000 francs for means of communication with other worlds, feeling perfectly sure that it will be awarded to me," he said. "The money, of course, is a trifling consideration, but for the great historical honor of being the first to achieve this miracle I would be almost willing to give my life."[2] Years later, however, the Institute of France denied that it had ever received an entry from Tesla. In fact, the Guzman Prize is still awaiting a successful claimant.

"My most important invention from a practical point of view," Tesla continued, "is a new form of tube with apparatus for its operation. In 1896 I brought out a high potential targetless tube which I operated successfully with potentials up to 4 million volts.... At a later period I managed to produce very much higher potentials up to 18 million volts, and then I encountered insurmountable difficulties which convinced me that it was necessary to invent an entirely different form of tube in order to carry out successfully certain ideas I had conceived. This task I found far more difficult than I had expected, not so much in the construction as in the operation of the tube. For many years I was baffled . . . although I made a steady slow progress. Finally . . . complete success. I produced a tube which it will be hard to improve further. It is of ideal simplicity, not subject to wear, and can be operated at any potential, however, high. . . . It will carry heavy currents, transform any amount of energy within practical limits, and it permits easy control and regulation of the same. I expect . . . results undreamed of before. Among others, it will [make possible] the production of cheap radium substitutes in any desired quantity and will be, in general, immensely more effective in the smashing of atoms and the transmutation of matter." He cautioned that it would not, however, open up a way to utilize atomic energy since his research had convinced him that this was nonexistent.[3]

He confessed to a certain annoyance because some newspapers had announced that he was prepared to give a full description of his remarkable tube. This would be impossible.

Because of "some obligations I have undertaken regarding the application of the tube for important purposes," he explained, "I am unable to make a complete disclosure now. But as soon as I am relieved of these obligations a technical description of the device and of all the apparatus will be given to scientific institutions."

No patents were ever filed nor was a prototype displayed. The second discovery he wanted to announce at his party consisted of "a new method and apparatus for the obtainment of vacua exceeding many times the highest heretofore realized. I think that as much as one-billionth of a micron can be attained. What may be accomplished by means of such vacua . . . will make possible the production of much more intense effects in electron tubes."*

There was a pause while wine was poured for his guests and glasses raised. Then the old man explained that he did not agree with ideas currently held regarding the electron. He believed that when an electron left an electrode of extremely high potential and in very high vacuum, it carried an electrostatic charge many times greater than normal.

"This may astonish some of those who think that the particle has the same charge in the tube and outside of it in the air," he said. "A beautiful and instructive experiment has been contrived by me showing that such is not the case, for as soon as the particle gets out into the atmosphere it becomes a blazing star owing to the escape of the excess charge. . . ."†

---

* This may have referred to improvements on the patent filed in 1922 which Tesla did not process to completion.

† Maurice Stahl suggests that the "blazing star" from Tesla's high-vacuum discharge tube may have been Leonard rays, which are very high-speed electrons able to penetrate very thin windows and show luminous paths by ionization of air molecules. This experiment does not necessarily multiply charged electrons. However, Tesla himself did not think this the effect he observed.

Tesla may have been on to something. Four decades later, the returns are still not in on the electrical charge of the electron. Physicists have been trying for years to calculate the charges of subatomic and larger particles. Despite confusing results, no one but Tesla had been willing to suggest that an electric charge could exist that was not equal to the charge of an electron, or of integral multiples thereof—no one, that is, until 1977 when three American physicists reported that they had discovered "evidence for fractional charge."

The result, if confirmed, "is likely to stand as one of the most important results in physics of this or any century," reported *Science News.*[4] Whether or not subparticles called "loose quarks" are involved in this esoteric mystery may prove to be at the heart of the matter. Tesla, although he did not know a quark from a gluon and lacked the elaborate research equipment of contemporary scientists, had at his service what Hobson had once described as his "cosmic intuition."

The eighty-first birthday party was a replay of the year before in terms of the inventions announced by the guest of honor, but it brought more international recognition.

His old friend Ambassador Konstantin Fotić presented the Grand Cordon of the White Eagle, the highest order of Yugoslavia, in behalf of young King Peter II through the Regent Paul. Then the Minister from Czechoslovakia, not to be outdone, presented to Tesla the Grand Cordon of the White Lion in the name of President Eduard Beneš. With this came an honorary degree from the University of Prague.

On this occasion the reporters questioned Tesla closely on his repeated claim of having perfected an interplanetary communication system. Once more he alluded to his intention of seeking the Pierre Guzman prize for this achievement.

The invention, he said, was "absolutely developed."

"I couldn't be any surer that I can transmit energy 100 miles than I am of the fact that I can transmit energy 1 million miles up," he said. He spoke of a "different kind of energy," as he had in the past, that would travel through a channel of less than one-half of one-millionth of a centimeter.[5]

Life on other planets was a "certitude." One problem that troubled him, he said, was the danger of hitting other planets with his "needle-point of tremendous energy," but he hoped that astronomers would help to solve this problem.

His point of energy, said the inventor, could easily be aimed at the moon, and Earthlings would then be able to see the effects, "the splash and the volatilization of matter." He suggested that advanced thinkers on other planets might even mistake the Tesla energy beam for some form of cosmic ray.

Once again he alluded to his atom-smashing electronic tube with which cheap radium could be produced. "I have built it, demonstrated, and used it. Only a little time will pass before I can give it to the world."

Were these merely the ramblings of an old man clinging to youthful dreams? The professors pooh-poohed them, but science writers as usual took him seriously. The world was on the verge of global war. William L. Laurence of *The New York Times* quoted Tesla in 1940 on the potential of erecting a "Chinese Wall" of his "teleforce" rays around the United States, which could melt airplanes at a distance of 250 miles. With $2 million to build a projection plant (was this the "limitless" market for steel Tesla had spoken of?), he claimed this could be done in three months. Laurence proposed that the government take him up on it. The War Department, as usual, made no overture to the inventor.

Teleforce, said Tesla, was based on four new inventions, of which two had already been tested: 1. a method of producing rays in the free air without a vacuum; 2. a method of producing "very great electrical force"; 3. a method of amplifying this force; and 4. a new method for producing "tremendous electrical propelling force."[6]

For years Tesla's biographers would be unable to find evidence to support the existence of working papers on these inventions. United States security agencies would consistently disavow knowledge of such matters; which was curious, because biographer O'Neill declared that federal agents removed from his home even nonsensitive papers of Tesla's, and he was never afterward able to discover who had actually "borrowed" his files.

Both O'Neill and (finally) Swezey were to conclude that Tesla's so-called secret weapons were "so much nonsense." O'Neill said, "The only knowledge I had was a firm belief that his theories, never adequately revealed to form a basis for judgment, were totally impracticable." At the same time, however, he admitted that he was never privy to any of Tesla's unpublished papers and that trying to get information from the inventor had always caused him to clam up in a direct ratio to the effort exerted.

A further curious fact was that even Tesla's proposals for his turbine and aircraft appeared to vanish from the federal archives.

One of the final honors that came to the inventor found him too ill to make a personal acceptance. The Institute of Immigrant Welfare in 1938 invited him to a ceremonial dinner at the Hotel Biltmore for a citation. His friend Dr. Rado read his speech, which contained high praise of George Westinghouse "to whom humanity owes an immense debt of gratitude." In absentia Tesla again claimed that he would win the Pierre Guzman prize for his work in cosmic communication.

His last years were not entirely fixed on outer space, however, nor were they even entirely cerebral. Some of his intellectual friends were surprised, even embarrassed, when with obvious pleasure, he began to fraternize with certain shy, burly, broken-nosed gentlemen of the boxing ring. This late-blooming fascination with pugilists and boxing confused both Swezey and O'Neill.

"Brain Dines Brawn," declared a wire photo caption. A happy Tesla was shown at table with the amiable Zivic brothers: "Dr. Nicola Tesla, famed inventor, broke his five-year, self-imposed exile in his suite at the Hotel New Yorker on Dec. 18th when he played host to Fritzie Zivic, welterweight champion. . . . Dr. Tesla, an ardent sports fan, predicted that Zivic would beat Lew Jenkins in their non-title bout . . ." The ever-admiring O'Neill, present at one of these meals, claimed that the psychic energy zinging between Tesla and the brothers made his own skin itch and tingle. Another writer present admitted feeling the same odd effects.

Removed as he was from the events in Europe, Tesla was not to be spared the tragedies of war in his last years. The honors conferred upon him by Yugoslavia and Czechoslovakia had been expressions by countries enjoying their last gasps of intellectual freedom. Hitler soon invaded Austria and his demands for the autonomy of Germans in the Sudetenland led to a crisis of government in Czechoslovakia. President Eduard Beneš resigned after Britain, France, and Italy, without even consulting his government, acceded to German occupation of the Sudetenland.

Next Yugoslavia's Regent Paul outraged the people of that country by agreeing to a compromise with Hitler that committed Slavs to join the Axis powers. For once diverse factions of Yugoslavia pulled together in defiance—Army, church, and peasants; Serbs, Croats, and Slovenes. Immediately, the pro-Allied Serbian military elements staged a successful coup and replaced Prince Paul with the seventeen-year-old King Peter II, who ascended the throne on March 28, 1941.

It pleased Tesla that the son of King Alexander, whom he had admired, was now the monarch. His closest friends in the New York/Washington Slavic communities remained those of "Great Serbian" outlook attached to the Yugoslav Embassy under Ambassador Fotić. At that time the only Croat on the Embassy staff was a young aide named Bogdan Raditsa (now a professor of Balkan history at Fairleigh Dickinson University). But soon Tesla's nephew, Sava Kosanović, a Serb born in Croatia, arrived in America to play what seemed to the frail old man a worrying and perplexing role.

Events began moving too fast. The inventor, aware mainly of tensions and shifting alliances among the local Slavic population, scarcely grasped the fact that as the greatest living hero of the Yugoslavs, he had been singled out by fate as an ideological pawn between East and West.

28. DEATH AND TRANSFIGURATION

The new government of King Peter, with broad popular support, confronted the Germans and refused to ratify the compromise agreement that had been made with Hitler by Prince Paul. Almost at once reprisals began.

On Palm Sunday, 1941, three hundred Luftwaffe bombers swept over the Yugoslav capital of Belgrade. Methodically they crisscrossed the city street by street, strafing everything that breathed. By noon 25,000 civilians were dead, and the wounded lay everywhere. Most public buildings were left in ruins, including the modern laboratory known as the Tesla Institute.

The combined armed forces of Germany, Italy, Hungary, and Bulgaria invaded the doomed country. Within only days the Yugoslav Army was crushed, and King Peter was sent to England for safety. His government-in-exile would operate from London for the remainder of World War II.

This, however, was only the beginning of the war for Yugoslavs. Accustomed to successive invasions for a thousand years, the people were resilient. The remnants of the Army and Communist factions withdrew into the mountains, from which they launched guerrilla attacks on the invaders. These armed fighters, men and women, were supplied with food grown by the old people and children remaining in undefended villages.

Against them the Nazis and Fascists carried out murderous reprisals. In the fishing villages and along the stony slopes of the Adriatic, half the people in every hamlet were systematically shot.

Soon, however, it became apparent to military strategists in the United States and England that, not only were Axis forces killing

Yugoslavs, but rival guerrilla factions of monarchists and Communists had begun to vie for Allied support and were shooting each other as well as the invaders.

Col. Drazha Mihailović, a Serbian army officer, led a faction called Chetniks (the "Yugoslav Army in the Fatherland"), composed mainly of Serbian and Bosnian monarchists. With close ties to King Peter, they became the first major resistance movement in Europe.[1] The initial British aid to Yugoslavia went to the Chetniks, but it was short-lived. The National Liberation Army or Partisans, led by Josip Broz Tito of the Communist Party, was swiftly rising to prominence.

Allied strategists knew little of Tito. It was said he had been left wounded on a battlefield in 1917 and captured by the Russians. There he was trained as a Communist leader and sent to France during the Spanish Civil War to aid the Loyalists or Republicans.

A Croat, Tito had little reason to love the monarchy, for he was imprisoned after returning to Yugoslavia. On release, he became active in organizing a metal workers' union and helped to build the Yugoslav labor movement. His emergence as head of the Partisans in World War II was that of a natural leader who inspired his fighters and maintained rigorous discipline. He was looking ahead to a time when the Slavs could rebuild a free and united country without oppression either by foreigners or kings.

Tito's goal was to set up committees of popular liberation after the Russian style, while Mihailović and the Chetniks favored local administrative authorities under the monarchy. Both factions kept on killing Germans and Italians but, unfortunately, they also continued murdering each other.[2]

Prof. Bogdan Raditsa,* then director of the information service of the Yugoslav Embassy in Washington, D.C., recalled, "The situation became rather complicated when Yugoslavia collapsed in 1941 and when, at the end of that year, a Royal Yugoslav Mission came to this country." It was com-

---

*Raditsa belonged to a family in southern Croatia that had always favored a union of Croatians and Serbs.

posed of members of King Peter's government and the Ban (Governor) of Croatia, Dr. Ivan Šubašić. Sava Kosanović, Tesla's nephew, then a member of the Democratic Party, also arrived as a minister of the exiled government.

"As soon as Kosanović came to the States," said Professor Raditsa, "he tried to reorient Tesla from the exclusive Serbian policy, and he succeeded. Tesla, even before, never felt himself a Great Serbian chauvinist. He used to say, 'I am a Serb but my fatherland is Croatia.'"[3]

The conflict between Serbs and Croats in exile intensified as the war went on, paralyzing normal Slav diplomatic activities in London, Washington, and New York.

"Kosanović, though a Serb," recalled Raditsa, "was leading the struggle for a brotherhood between the Serbs and Croats against Fotić and many other Serb members of various Yugoslav missions. Thus he began using Tesla for the policy directed against the Great Serbians.

"Tesla himself . . . was not aware of the deep conflict between the Serbs and Croats, and as basically a scientist and in old age, he was very candid in politics."

Raditsa said he seemed happy that he finally had a man of his own blood near him in New York and noted that Tesla began to rely upon Kosanović's opinion on everything. During this period the inventor was receiving about $500 per month from the royal government as an honorarium.

Various political messages elicited from Tesla for home consumption, said Raditsa, were actually written by Kosanović.[4]

Toward the end of 1942 the Yugoslav Information Center was opened in New York in the Royal Mission headquarters on Fifth Avenue. Raditsa and Kosanović worked together at this office, issuing bulletins and other publications. But a crisis broke out when news reached them of the fighting between Mihailović and Tito.

"Kosanović," Raditsa said, "joined Tito and began to popularize the National Liberation Movement for a new Yugoslavia. He had a terrible time to convince Tesla that monarchy was losing in Yugoslavia and that a new Yugoslavia was beginning to come out from the fratricidal civil war.

As the largest majority of Serbs in Croatia were joining Tito, Kosanović convinced Tesla that he too should join the movement that was largely shared by the masses of the people, Serbs and Croats. So Tesla's message to the Serbs and Croats was written by Kosanović."[5]

On the walls of the Tesla Museum in Belgrade one may read a vastly enlarged photocopy of the words allegedly sent by Tesla to his embattled countrymen only months before his own death. American Vice-President Henry A. Wallace also had a hand in its drafting. Typewritten, it has many cross-outs and interlinings in Tesla's own handwriting, yet the style is that of an ideologue, which the inventor was not:

> Out of this war . . . a new world must be born, a world that
> would justify the sacrifices offered by humanity. This . . . must
> be a world in which there shall be no exploitation of the weak
> by the strong, of the good by the evil, where there will be no
> humiliation of the poor by the violence of the rich; where the
> products of the intellect, science, and art will serve society for
> the betterment and beautification of life, and not the individuals
> for achieving wealth. This new world shall not be a world of the
> downtrodden and humiliated, but of free men and free nations,
> equal in dignity and respect for man.

The inventor's name also appeared on another message—sent to the Soviet Academy of Sciences on October 12, 1941, urging joint struggle against the Axis powers by Russia, Great Britain, and America, in aid of the revolutionary struggle of the Yugoslav people. This message is not to be seen in the Museum, however, presumably because nostalgia Russian-style has ceased to be politic.

Kosanović became chairman of the Yugoslav Economic Mission advocating a New Yugoslav federation versus the centralistic prewar royalist Yugoslavia. This new organization also began working for a new Central East European Federation. Raditsa too became a member of the Tito movement.

King Peter was desperately seeking for Mihailović the support of President Franklin Delano Roosevelt and Prime Minister Winston Churchill, as well as that of his own Uncle Bertie, who was King George VI of England. The British, at first sympathetic to the Chetnik cause, began to change as they received reports of the aggressive actions of Tito's Partisans.

In 1942 King Peter visited Washington to intercede with FDR. Yugoslav pilots were being trained in Tennessee. FDR told him that America would send airplanes to the Chetniks as soon as they could be spared from the war in the Middle East. The monarch visited New York City, attending a large reception for the American Friends of Yugoslavia at the Colony Club. The Colony, the first female socialites' club in America, had been founded at the inspiration of energetic Anne Morgan. She attended the function, as did the King's mother, Queen Marie, and Mrs. Roosevelt. It was the sort of affair Tesla himself would have delighted in had he not been weak and ill. So King Peter went to him.

In his diaries (*A King's Heritage*), under date July 8, 1942, the young Peter II wrote: "I visited Dr. Nicola Tesla, the world-famous Yugoslav-American scientist, in his apartment in the Hotel New Yorker. After I had greeted him the aged scientist said: 'It is my greatest honor. I am glad you are in your youth, and I am content that you will be a great ruler. I believe I will live until you come back to a free Yugoslavia. From your father you have received his last words: "Guard Yugoslavia." I am proud to be a Serbian and a Yugoslav. Our people cannot perish. Preserve the unity of all Yugoslavs—the Serbs, the Croats, and Slovenes.'"

The King added that he was deeply touched and that both he and Dr. Tesla had wept. He then visited Columbia University, to be warmly welcomed by President Nicholas Murray Butler and to find another link with his own country in the Pupin Physics Laboratory.

Returning to Washington, he was assured by FDR that food, clothing, arms, and ammunition would be dropped over Yugoslavia. But he was shocked when, in 1943, the British Mission in Yugoslavia made official contact with Tito. Peter asked to be parachuted into his country, but Churchill demurred. Tito openly accused Mihailović of being a traitor.[6]

At the Teheran Conference in November there occurred, largely at Churchill's instance, what the King described as a "fatal change" of Allied policy. It was decided that "the basic force fighting the Germans in Yugoslavia recognized by the Allies was the National Liberation Army, under the command of Tito, and the Partisan force received full recognition as an Allied Army. Mihailović was thus denied and abandoned."[7]

Winston Churchill overnight became a hero of modern Yugoslavia. And when the young monarch frantically wrote to FDR for support, the ailing President replied urging him to accept Churchill's advice "as if it was my own." Within months Roosevelt died.

Tesla's nephew Kosanović, along with certain other diplomatic representatives of King Peter, had been dismissed by the monarch at the height of the 1942 crisis. He often told Bogdan Raditsa in those days that he felt Tesla had been terribly shocked by his nephew's exclusion from the royal government. In fact, Kosanović believed that the inventor's death was actually precipitated by his own "setback."

"He thought," Kosanović repeatedly told Raditsa, "that I was punished, and that eventually I would be arrested or something of the kind, but I succeeded to convince him that it was inevitable in politics."[8]

During this period Kosanović was frank in saying that he tried to keep Tesla from seeing members of the royal government. Ambassador Fotić had become "the enemy" since he still favored a Great Serbian policy as opposed to the changes ahead. Tesla's relationship with this old friend became "lukewarm."

"There is no doubt," said Professor Raditsa, "that the whole internecine tragedy of Yugoslavia from 1941 to 1943 must have had a rather depressing impact upon Tesla. Very often he would ask me, could I explain to him what was going on among us, and why we cannot agree...."

After the war, Mihailović would be executed by a "People's Court" for alleged collaboration with the enemy, and the Republic of Yugoslavia declared to exist, with Tito as President for life and the Communists firmly in charge.

A count of Yugoslavian casualties at the end of World War II disclosed that 2 million persons had died; tragically, many thousands had been killed by fellow Yugoslavs.

"After the war," recalled Professor Raditsa, "Kosanović became a minister in the Tito-Šubašić Government, and I was his assistant in the Ministry of Information from 1944 to 1945, when I left the country, for I couldn't become a Communist. Later on in 1946, Sava Kosanović became Tito's ambassador in Washington but I never saw him again after I left Belgrade in October of 1945. Kosanović had accepted totally the Communist system in Yugoslavia and remained loyal until his death."

There had not been a time in ten centuries when the Yugoslavs had not been ruled and ransacked by invaders—by Venetians, Romans, Turks, Bulgars, Austrians, Hungarians, Germans, Italians, when they were not living under threat of torture, prison, or violent death. Now a marvelous truth began to dawn upon them: that they were free, in a manner of speaking.

Tesla would not live to see this. Whether he could ever have accepted the new government, with its Soviet-type Constitution and a Soviet alliance, whether he could ever have accepted the permanent exile of his beloved monarch, are unanswerable questions.

Unfortunately, however, all this was to have a bearing on how he would be remembered in the West. The fading of his scientific reputation, the forgetfulness of Americans in the postwar period, resulted in large degree from the disappearance of most of his scientific papers behind that new Cold War phenomenon, the Iron Curtain.

In 1948 Yugoslavia ceased to be an Iron Curtain country, declaring its independence from the Soviet doctrine of "limited sovereignty." America and her allies then were generous in sending economic and military aid to the Slavs; but the damage had been done. America had not raced to Tito's wartime support with the alacrity that Churchill had shown. In the future it would not be made easy for American scholars to draw on Yugoslav sources to document the achievements of Nikola Tesla.

The inventor became very feeble in the winter of 1942. His fear of germs was so obsessive that even his closest friends were required to stand at a distance, like the subjects of a neurotic Tudor. (Pigeon germs did not seem to worry him.) He had heart trouble and suffered occasionally from fainting spells. No longer able to feed his beloved birds, he often relied upon a young man named Charles Hausler, who owned racing pigeons, to take care of them for him.

Hausler had worked for Tesla in this capacity from around 1928 onward, his job being to go to the New York Public Library at noon each day with grain and then to walk around the four sides of the building looking for young or injured birds on window sills or behind large statues. He would take them to Tesla's hotel for rest and recuperation. Then, he has recalled, "I would release them at the library for him." He remembered that the cages in Tesla's rooms had been built by a fine carpenter—"as Mr. Tesla was in all his doings it had to be done right." The pigeons also enjoyed a curtained shower bath.

Hausler and Tesla spent many hours together, talking mostly of pigeons. Once Tesla confided to him that "Thomas Edison could not be trusted." The boy remembered his employer as "a very kind and considerate human person," and there was one incident that stood out in his mind long afterward. "He had a large box or container in his room near the pigeon cages and he told me to be very careful not to disturb the box," said Hausler, "as it contained something that could destroy an airplane in the sky and he had hopes of presenting it to the world." He believed it probably was stored in the cellar of the hotel later.[9]

On a bitter day in early January 1943, Tesla called his other messenger boy, Kerrigan, and gave him a sealed envelope addressed to Mr. Samuel Clemens, 35 South Fifth Avenue, New York City. The boy set forth into the whipping wind and searched fruitlessly for the number. As it turned out, this had been the address of Tesla's first laboratory; but now South Fifth Avenue was West Broadway, and no one by the name of Samuel Clemens lived in the area.

Kerrigan made his way back to the Hotel New Yorker and reported to the sick man. In a weak voice, Tesla explained that Clemens was the famous Mark Twain and that everyone knew of him. He sent Kerrigan forth once more, and this time asked him also to take care of the pigeons. The perturbed messenger fed the birds and then consulted his supervisor, who told him that Mark Twain had been dead for twenty-five years. Once again Kerrigan trudged through the cold afternoon to Tesla's rooms, where he explained and tried to return the envelope.

The inventor was indignant and refused to hear that the humorist was dead. "He was in my room last night," he said. "He sat in that chair and talked to me for an hour. He is having financial difficulties and needs my help. So—don't come back until you have delivered that envelope." Once again the messenger went to his supervisor and together they opened the envelope. It contained a blank sheet of paper wrapped around twenty five-dollar bills—enough to help an old friend through a little fainting spell.

On the fourth of January, the inventor, although very weak, went to his office to make an experiment that George Scherff was interested in. Scherff dropped in to help him prepare for it. The work was interrupted, however, when Tesla felt a recurrence of some sharp pains in his chest.

Refusing medical aid, he returned to his hotel. Next day a maid came in and cleaned. As she left, he asked her to put the *Do Not Disturb* sign on his door to keep visitors away, and not to bother cleaning. The sign remained there the following day and the one after that.

Early on the morning of January 8, Alice Monaghan, a maid, ignored the sign and entered the apartment to find the inventor dead in bed, his sunken, emaciated face composed.[10] Assistant Medical Examiner H. W. Wembly examined the body, placed the time of death as 10:30 P.M. on January 7, 1943, and gave his opinion that the cause of death had been coronary thrombosis. Tesla had died in his sleep, and the examiner noted that he had found "No suspicious circumstances." The inventor was eighty-six years of age.

Kenneth Swezey was notified at once; and at ten o'clock that morning he telephoned to Dr. Rado at New York University. King Peter's headquarters, then at 745 Fifth Avenue, was advised by the professor. Tesla's nephew Kosanović, then wartime president of the Eastern and Central European Planning Board for the Balkan countries, also was notified.

Then the FBI was called. Swezey and Kosanović summoned a locksmith and Tesla's safe was opened and the contents examined.

The body was removed to the Frank E. Campbell Funeral Home at Madison Avenue and 81st Street and a sculptor was engaged by Hugo Gernsback to prepare a death mask of the inventor.

Just before Tesla's death, Eleanor Roosevelt had tried to intercede in his behalf with President Roosevelt—perhaps with the idea of conferring some honor upon him. In the Tesla Museum at Belgrade three brief notes on White House stationery may be read. On January 1, at the request of author Louis Adamić, Mrs. Roosevelt had promised to ask the President to write to Tesla and said that she herself would call on him on her next trip to New York. The second note is headed, "Memo for Mrs. Roosevelt" and is signed FDR: "I was having this looked into but the papers yesterday carried the story that Dr. Tesla had died. Therefore I am returning the enclosures herewith." A third note of January 11 from Eleanor Roosevelt to Admić forwards the President's message and adds her sorrow at learning of the inventor's death.

Adamić wrote a moving eulogy to Tesla that was read by New York Mayor Fiorello H. LaGuardia over station WNYC on January 10.[11] Meanwhile the extreme tensions between Serb and Croat factions in the United States were making the planning of funeral services difficult. The body lay in state but, according to an unpublished letter of O'Neill's, "only twelve people, some of whom were newspaper reporters," came to view it.

When state services were held at four o'clock on January 12, in the Cathedral of St. John the Divine, however, more than two thousand people crowded in. Serbs and Croats were seated on opposing sides of the cathedral, Bishop William T. Manning having exacted from both factions a promise of no political speeches. The service was begun in English by Bishop Manning and concluded in Serbian by the Very Rev. Dusan Sukletović.

Among Balkan diplomats present were Ambassador Fotić, the Governor of Croatia, a former Prime Minister of Yugoslavia, and the Minister of Food and Reconstruction. In the front row with Kosanović, chief mourner and head of the important new trade mission, sat Swezey. Dr. Rado had been too ill to attend as an honorary pallbearer.

Figures important in American science and industry who did attend as honorary pallbearers included Professor Edwin H. Armstrong, Dr. E. F. W. Alexanderson of General Electric, Dr. Harvey Rentschler of Westinghouse, engineer Gano Dunn, and W. H. Barton, curator of the Hayden Planetarium of the American Museum of Natural History. Newbold Morris, president of the New York City Council, headed this group.

When word of Tesla's death spread abroad to war-stricken Europe, telegrams of tribute and sorrow began pouring in from scientists and governmental leaders alike. In the United States three Nobel prizewinners in physics, Millikan, Compton, and James Franck, joined in a eulogy to the inventor as "one of the outstanding intellects of the world who paved the way for many of the important technological developments of modern times."

The President and Mrs. Roosevelt expressed their gratitude for Tesla's contributions "to science and industry and to this country." Vice-President Wallace, in the spirit of the new Yugoslavia, declared that, "In Nikola Tesla's death the common man loses one of his best friends."

Although Louis Adamić wrongly eulogized Tesla as one who had cared nothing for money, he could not have been more accurate when he said that Tesla was not really dead: "The real, important part of Tesla lives in his achievement, which is great, almost beyond calculation, and an integral part of our civilization, our daily lives, our current war effort.... His life is a triumph...."[12]

Among the honors that had come to Tesla in his life were many academic degrees from American and foreign universities; the John Scott Medal, the Edison Medal, and various awards from European governments. In September 1943 the Liberty ship *Nikola Tesla* was launched, an honor that would have pleased the scientist. But not until 1975 was he inducted into the National Inventors Hall of Fame.

Eight months after Tesla's death, the U.S. Supreme Court handed down the decision that he had been confident would come eventually—ruling that he was the inventor of radio.

His body was taken to Ferncliffe Cemetery at Ardsley-on-the-Hudson in the deep cold of the winter afternoon. In the car that followed the hearse rode Swezey and Kosanović. The inventor's remains were cremated and his ashes later returned to the land of his birth.*

In almost every nation in the world, the fighting and dying continued.

---

* Charlotte Muzar, formerly secretary to Sava N. Kosanović, carried Tesla's ashes to the Tesla Museum in Belgrade in 1957. Throughout the years Kosanvić had spoken of leaving the ashes in America and had hoped an appropriate memorial to the inventor would be raised in the United States as their resting place.
—Archives, Tesla Memorial Society.

29. THE MISSING PAPERS

In addition to his acknowledged achievements, Tesla left a legacy of riddles. To pose only three of the most major: Was his unrealized concept for the wireless transmission of energy through the Earth scientifically valid? What actually was he doing in his experimentation with death/disintegrator beam weapons? And what became of his unpatented research papers and other sensitive documents in the days immediately following his death?

In the category of subquestions, what turn of affairs rekindled the intense interest of the U.S. intelligence establishment in Tesla's work (as something surely did) in the late 1940s?

Like Einstein he had been an outsider and, like Edison, a wide-ranging generalist. As he himself had said, he had the "boldness of ignorance." Where others stopped short, aware of what could not be done, he continued. The survival of such mutants and polymaths as Tesla tends to be discouraged by modern scientific guilds. Whether either he or Edison could have flourished in today's milieu is conjectural.

The example set by Tesla has always been particularly inspiring to the lone runner. At the same time, however, his legacy to establishment science is profound for his research, although sometimes esoteric, was almost always sweeping in its potential to transform society. His contribution was major rather than incremental. His turbine failed in part because it would have required fundamental changes by whole industries. Alternating current triumphed only after it had overcome the resistance of an entire industry.

But there was an unfortunate corollary to Tesla's lone battles with the scientific-industrial establishment. Since he was part of no group or institution, he had no colleagues with whom to discuss work in progress, no formal, accessible repository for his research notes and papers. He worked not just in private, but—his love of flamboyant announcements to the press notwithstanding—in secret. Thus any inventions which he did not patent or give freely to the world were more or less shrouded in mystery. And, because of the handling of the papers he left behind after his death, the range of his achievement continues to remain a partial mystery.

If this has been frustrating to the scientists who have succeeded Tesla, it has at least been stimulating. After a period of obscurity, the one hundredth anniversary of his birth in July 1956 brought an international reawakening to the importance of the inventor's life and genius. Interest in his work, fired by a growing awareness of the riddles surrounding it, has been escalating ever since, almost as if he had been reborn in his true psychological age.

He was honored by centennial celebrations in America and Europe. The American Institute of Electrical Engineers dedicated its fall meeting in Chicago to a review of his life and inventions. Commemorative programs were arranged by the Institute of Radio Engineers, the Chicago Museum of Science and Industry, the Franklin Institute, and various universities, the Tesla Society playing an active role in such recognition. Permanent memorials in the form of scholarships and medals were proposed and exhibits presented by science museums. Special ceremonies were conducted at Niagara Falls, and a statue was later erected in his honor on Goat Island, a gift from the people of Yugoslavia. Chicago, reminded by attorney/author Elmer Gertz that it should be eternally grateful to him for having made the Columbian Exposition of 1893 the "wonder of the globe," dedicated a new public school to Tesla's memory.

The inventor's old colleagues of the AIEE journeyed to Europe to attend more celebrations, statue unveilings, and dedications in his honor. The International Electrotechnical Commission in Munich took formal action, making his name an international scientific unit, the *tesla* joining such historic electrical symbols as farad, volt, ampere, and ohm.[1]

As the exploration of space accelerated, so did interest in Tesla, especially from the standpoint of beam weaponry and microwave work. In America, Russia, Canada, and various other countries, projects in his name or derived from his pioneering, from weather-control to nuclear fusion, began to attract scientific attention. Some were just the shoestring efforts of loners, their laboratories old Quonset huts. Some were top secret and financed by enormous budgets.

Tesla's year of secret experiments at Colorado Springs in 1899 provided the basic impetus for much of this new exploration. His *Colorado Springs Notes*, when they appeared in English in 1978 under the imprint of the Tesla Museum at Belgrade, were eagerly awaited by many scientists. But even this work left important questions unanswered.

The bulk of his papers having vanished from America, reliable information was harder to come by than the recurring rumors of conspiracy, espionage, and patent theft. Scientists thought it strange that some aspects of his Colorado Springs research found in scattered sources did not appear in the Yugoslav-published *Notes*. Only by piecing together fragmentary information could the magnitude of his experiments be comprehended.

Around 1928 O'Neill, by merest chance, had happened to see a legal advertisement in a New York newspaper announcing that six boxes placed in storage by Nikola Tesla would be sold by the storage warehouse for unpaid bills. Feeling that such material should be preserved, he went to the inventor and asked permission to try to obtain funds to reclaim the material.

"Tesla hit the ceiling," he recalled. "He assured me he was well able to take care of his own affairs. . . . He forbid me to buy them or do anything in any way about them."

Shortly after the inventor died, O'Neill got in touch with Sava Kosanović, told him about the boxes, and urged him to protect them. He was never able to get a positive statement from Kosanović that he had obtained the boxes and examined the contents. "He gave evasive assurances that there was no reason for me to worry. . . ."

Others too were interested in the papers. A young American engineer engaged in war work consulted Tesla on a ballistics engineering problem because he could not get time on an overworked computer, and Tesla's mind was known to offer the nearest thing to it. Soon he became fascinated with Tesla's scientific papers and was allowed to take batches of them home to his hotel room where he and another American engineer pored over them each night. They were returned the next day, a procedure which continued for about two weeks prior to the inventor's death.

Tesla had received offers to work for Germany and Russia. After the inventor died, both engineers became concerned that critical scientific information might fall into foreign hands and alerted United States security agencies and high government officials.

The relevant records that I have obtained from federal agencies under the Freedom of Information Act reveal strange twistings and inconsistencies in the handling of the inventor's estate. Tesla left tons of papers, barrels and boxes full of them. But he left no will. He was survived by five nieces and nephews, of whom two lived in America at the time of his death.

Curiously, the FBI released his estate to the Office of Alien Property, which promptly sealed the contents. Since Tesla was an American citizen, the OAP's concern in the matter was hard to justify. After a court hearing, however, the estate was released to Ambassador Kosanović, one of the heirs.

Swezey, who also had hoped to write a biography of Tesla (his death intervened), received the following account in 1963 from a former aide of Ambassador Kosanović's:

"Back in 1943 ... when Tesla died, it was a matter of very short time when Mr. K was issued a certificate from or by the Office of Custodian of Alien Property conveying to Mr. K full rights to the Tesla papers.... he had them all packed up and sent off to the Manhattan Storage Company where they remained until ready for packing and shipping off to Yugoslavia in 1952. Mr. K paid for storage charges.... All this time the certificate from the Alien Property Office was in my possession (in case of need)....

"You will perhaps remember that a number of times Mr. K mentioned the fact that the custodian at the storage warehouse told him that some government guys were in to microfilm some of the papers. . . . when we opened the safe in the present museum building (in Belgrade, Yugoslavia) the bunch of keys, which was the last thing Mr. K. flung into the safe at the New Yorker Hotel before the combination was re-set to a new combination, were not found in the safe, but in an entirely different box. Also the gold medal (the Edison Medal) was missing from the safe. . . . Anyway, for years and years Mr. K was bothered by the fact that Tesla papers had been gone thru and just before his departure from Washington in 1949–50(?) he decided to follow my suggestion to call Edgar J. Hoover [sic] and ask him. Mr. Hoover denied categorically that the FBI had gone into the papers. . . ."

The aide said Tesla had told his nephew that "he wished to leave his works, property, etc., to his native country." (Not only is this uncorroborated but the papers were in English.)

Immediately after Tesla's death an exchange of telegrams flew between FBI Agent Foxworth of the field division of the New York Bureau and the director of the New York Bureau of the FBI. The day following discovery of the body, Agent Foxworth reported:

"Experiments and research of Nikola Tesla, deceased. Espionage—M. Nikola Tesla, one of the world's outstanding scientists in the electrical field, died January seventh, nineteen forty-three at the Hotel New Yorker, New York City. During his lifetime, he conducted many experiments in connection with the wireless transmission of electrical power and . . . what is commonly called the death ray. According to information furnished by X [name deleted], New York City, the notes and records of Tesla's experiments and formulae together with designs of machinery . . . are among Tesla's personal effects, and no steps have been taken to preserve them or to keep them from falling into hands of people . . . unfriendly to the war effort of the United Nations. ! . . ." (The FBI was, however, advised by the office of Vice-President Henry A. Wallace that the government was "vitally interested" in preserving Tesla's papers.)

Bloyce D. Fitzgerald, "an electrical engineer who had been quite close to Tesla during his lifetime," continued Foxworth, "advised the New York office that on January seventh, nineteen forty-three, Sava Kosanović, George Clark, who is in charge of the museum and laboratory for RCA, and Kenneth Swezey . . . went to Tesla's rooms in the New Yorker [author's note: the correct date would have been January 8], and with the assistance of a locksmith broke into a safe which Tesla had in his rooms in which he kept some of his valuable papers. . . . Within the last month, Tesla told Fitzgerald that his experiments in connection with the wireless transmission of electrical power had been completed and perfected.

"Fitzgerald also knows that Tesla had conceived and designed a revolutionary type of torpedo which is not presently in use by any of the nations. It is Fitzgerald's belief that this design has not been made available to any nation up to the present time. From statements made to Fitzgerald by Tesla, he knows that the complete plans, specifications and explanation of the basic theories of these things are some place in the personal effects of Tesla. He also knows there is a working model of Tesla's, which cost more than ten thousand dollars to build, in a safety deposit box belonging to Tesla at the Governor Clinton Hotel, and Fitzgerald believes this model has to do with the so-called death ray or the wireless transmission of electrical current.

"Tesla has also told Fitzgerald in past conversations that he has some eighty trunks in different places containing transcripts and plans having to do with experiments conducted by him. Bureau is requested to advise immediately what, if any, action should be taken concerning this matter by the New York Field Division."[2]

Kosanović later reported to Walter Gorsuch of the Office of Alien Property in New York that he first went to Tesla's rooms with the other men to search for a will. After the safe was opened, Swezey took from it a book containing the testimonials sent to Tesla on his seventy-fifth birthday, while Kosanović took from the room three pictures of Tesla. According to the manager of the New Yorker Hotel and Kosanović, nothing else was removed. The safe was closed under a new combination, which combination was in Kosanović's exclusive possession.

On January 9, Gorsuch of OAP and Fitzgerald went to the New Yorker Hotel and seized all of Tesla's property, consisting of about two truckloads of material, sealed it and transferred it to the Manhattan Storage and Warehouse Company. It was added to about thirty barrels and bundles that had been there since about 1934, and these too were sealed under orders of the OAP.

In addition to the question of the legitimacy of Alien Property's involvement in the case is the question of why Kosanović was allowed to have access to the safe's combination, from which he later claimed the Edison Medal had vanished. Tesla's American naturalization papers, which he so prized that he always kept them in his safe, may now be seen at the Tesla Museum in Belgrade; but it is not known what other papers or objects were in the safe.

The Washington Bureau of the FBI went so far as to advise the New York Bureau "to discreetly take the matter up with the State's Attorney in New York City with the view to possibly taking Kosanovich into custody on a burglary charge and obtaining the various papers which Kosanovich is reported to have taken from Tesla's safe." New York was also told to contact the Surrogate Court so stops could be placed against all of Tesla's effects, so that no one could enter them without an FBI agent being present, and New York was to keep Washington advised of all developments.[3]

The idea of arresting the Yugoslav ambassador was quickly dropped. And very soon the Washington headquarters made a curious decision. Edward A. Tamm of the FBI in Washington advised D. M. Ladd of that Bureau that the whole matter was being turned over to the Custodian of Alien Property; and Tamm noted, "There appears to be no need for us to mess around in it."[4]

Soon the well-known electrical engineer Dr. John G. Trump, who was serving as a technical aide to the National Defense Research Committee of the Office of Scientific Research and Development, was asked to participate in an examination of Tesla's scientific papers. Present at the Manhattan Warehouse & Storage Company in addition to Dr. Trump were Willis George, Office of Naval Intelligence, Third Naval District, Edward Palmer, chief yeoman, USNR, and John J. Corbett, chief yeoman, USNR.

Dr. Trump reported afterward that no examination was made of the vast amount of Tesla's property that had been in the basement of the New Yorker Hotel for ten years prior to his death, or of any of his papers except those in his immediate possession at the time of death. It should be remembered that Tesla's scientific reputation had been in eclipse for a number of years and that there had been many efforts to discredit his claims in radio, robotry, and alternating current. Dr. Trump was a busy man, just as the staff of the FBI was stretched thin by its preoccupation with investigating wartime sabotage.

"As a result of this examination," wrote Dr. Trump, "it is my considered opinion that there exist among Dr. Tesla's papers and possessions no scientific notes, descriptions of hitherto unrevealed methods or devices, or actual apparatus which could be of significant value to this country or which would constitute a hazard in unfriendly hands. I can therefore see no technical or military reason why further custody of the property should be retained."

He added: "For your records, there has been removed to your office a file of various written material by Dr. Tesla which covers typically and fairly completely the ideas with which he was concerned during his later years. These documents are enumerated and briefly abstracted in the attachment to this letter."

In closing Dr. Trump said: "It should be no discredit to this distinguished engineer and scientist, whose solid contributions to the electrical art were made at the beginning of the present century, to report that his thoughts and efforts during at least the past fifteen years were primarily of a speculative, philosophical, and somewhat promotional character—often concerned with the production and wireless transmission of power—but did not include new sound, workable principles or methods for realizing such results."

The file (of which Dr. Trump's notes were only an abstract) consisted apparently of either photostats or microfilm made by the naval officers present, and the original papers apparently remained in storage, later to be transmitted to Yugoslavia. The examination had failed to disclose any alien-owned property subject to the vesting power of the Alien Property Custodian under the Trading with the Enemy Act. Tesla's papers and personal effects were released in February of 1943 for disposition by Kosanović, the administrator of his estate.

Dr. Trump's abstract included the following:

"*Art of Telegeodynamics, or Art of Producing Terrestrial Motions at Distance*—This document, in the form of a letter dated June 12, 1940, to the Westinghouse Electric & Manufacturing Co., proposes a method for the transmission of large amounts of power over vast distances by means of mechanical vibrations of the earth's crust. The source of power is a mechanical or electromechanical device bolted to some rocky protuberance and imparting power at a resonance frequency of the earth's crust. The proposed scheme appears to be completely visionary and unworkable. Westinghouse's reply indicates their polite rejection....

"*New Art of Projecting Concentrated Non-Dispersive Energy through Natural Media*—This undated document by Tesla describes an electrostatic method of producing very high voltages and capable of very great power. This generator is used to accelerate charged particles, presumably electrons. Such a beam of high-energy electrons passing through air is the 'concentrated nondispersive' means by which energy is transmitted through natural media. As a component of this apparatus there is described an open-ended vacuum tube within which the electrons are first accelerated.

"The proposed scheme bears some relation to present means for producing high-energy cathode rays by the cooperative use of a high-voltage electrostatic generator and an evacuated electron acceleration tube. It is well known, however, that such devices, while of scientific and medical interest, are incapable of the transmission of large amounts of power in nondispersed beams over long distances. Tesla's disclosures in this memorandum would not enable the construction of workable combinations of generator and tube even of limited power, though the general elements of such a combination are succinctly described.

"*A Method of Producing Powerful Radiations*—an undated memorandum in Tesla's handwriting describing 'a new process of generating powerful rays or radiations.' This memorandum reviews the works of Lenard and Crookes, describes Tesla's work on the production of high voltages, and finally in the last paragraph gives the only description of the invention contained in the memorandum. . . . 'Briefly stated, my new simplified process of generating powerful rays consists in creating through the medium of a high-speed jet of suitable fluid a vacuous space around a terminal of a circuit and supplying the same with currents of the required tension and volume.'"

Long afterward in a letter to a colleague, Dr. Trump told what happened when he visited the Hotel Governor Clinton to examine the "device" stored in its vault, presumably the same box remembered by the messenger boy in Tesla's room.

"Tesla had warned the management that this 'device' was a secret weapon," said Dr. Trump, "and it would detonate if opened by an unauthorized person. Upon opening the vault and indicating the package containing the secret weapon, the hotel manager and employees promptly left the scene." The federal agents who had come along also pulled back, the better to give him the sole distinction of opening the parcel.

It was wrapped in brown paper and tied with a string. He remembered hesitating, thinking how beautiful the weather was outdoors, and pondering on why he was not outside too.

He lifted the parcel onto a table and, mustering his courage, snipped the string with his pocket knife. He removed the wrapping. Inside was a handsome polished wooden chest bound with brass. It required a final effort of courage to raise the hinged lid.

Inside stood a multidecade resistance box of the type used for Wheatstone bridge resistance measurements—a common standard item to be found in every electrical laboratory before the turn of the century!

Why had Tesla seen fit to terrify the staff and management of the hotel with this harmless object for so many years? Perhaps he had become so accustomed to having his hotel bills paid behind his back (believing that the hotels, honored to have him living there, had routinely dismissed the billings), that he was insulted when the Governor Clinton brashly demanded its $400.

Although the FBI closed its Tesla file in 1943, it didn't seem to want to stay closed. It was reopened in 1957 when an informant complained that a New York couple were issuing newsletters containing "information pertaining to flying saucers and interplanetary matters" and exploiting the inventor's name and fame. They allegedly claimed that Tesla's engineers, after his death, had completed a "Tesla Set," a radio device for interplanetary communication, that the device had been placed in operation in 1950 and since then Tesla engineers had been in close touch with alien spaceships. Once again the FBI decided no action was warranted and the file was closed.

Swezey had never put much credence in the "secret weapon" rumors and had written to an inquirer: "Because Tesla was a recluse, and himself liked to talk in mystifying terms during his later years, I think many legends have been built up about the dozens of ideas he had evolved but which were not permitted by others to see the light of day."

He said he had known the inventor well for two decades before his death: "Tesla's greatest genius flamed up during a dozen or so years just before and slightly after the turn of the century. What he did after that may have carried the germs of some of the developments we are witnessing today, but he had not carried any of them—at least on paper or in any other tangible form—to the point of practicality. . . ."

Perhaps, but between 1945 and 1947 an interesting exchange of letters and cables occurred among the Air Technical Service Command at Wright Field, Ohio, in whose Equipment Laboratory much top-secret research was being performed, Military Intelligence in Washington, and the Office of Alien Property—subject, files of the late Nikola Tesla.

On August 21, 1945, the Air Technical Service Command requested permission from the commanding general of the U.S. Army Air Force in Washington, D.C., for Private Bloyce D. Fitzgerald to go to Washington for a period of seven days "for the purpose of securing property clearance on enemy impounded property."

On September 5, 1945, Colonel Holliday of the Equipment Laboratory, Propulsion and Accessories Subdivision, wrote to Lloyd L. Shaulis of the OAP in Washington, confirming a conversation with Fitzgerald and asking for photostatic copies of the exhibits annotated by Trump from the estate of Tesla. It was stated that the material would be used "in connection with projects for National Defense by this department," and that all of it would be returned in a reasonable length of time.

That was the last time that the Office of Alien Property or any other federal agency in the United States admitted to having possession of Tesla's papers on beam weaponry. Shaulis wrote to Colonel Holliday on September 11, 1945, saying, "The materials requested have been forwarded to Air Technical Service Command in care of Lt. Robert E. Houle. These data are made available to the Army Air Force by this office for use in experiments; please return them." They were never returned.

These were the full photostatic copies, not merely the abstracts. OAP has no record of *how many* copies were made by those who examined the files with Dr. Trump. The Navy has no record of Tesla's papers; *no* federal archives have a record of them.

Curiously, four months after the photostats had been sent to Wright Field, Col. Ralph Doty, the chief of Military Intelligence in Washington wrote James Markham of Alien Property indicating that they had never been received: "This office is in receipt of a communication from Headquarters, Air Technical Service Command, Wright Field, requesting that we ascertain the whereabouts of the files of the late scientist, Dr. Nichola [sic] Tesla, which may contain data of great value to the above Headquarters. It has been indicated that your office might have these files in custody. If this is true, we would like to request your consent for a representative of the Air Technical Service Command to review them. In view of the extreme importance of these files to the above command, we would like to request that we be advised of *any attempt by any other agency to obtain them*. [Italics supplied.]

"Because of the urgency of this matter, this communication will be delivered to you by a Liaison Officer of this office in the hope of expediting the solicited information."

The "other" agency that had the files, or should have had them, was the Air Technical Service Command itself! Colonel Doty's letter, which was classified under the Espionage Act, was declassified on May 8, 1980.

This embarrassing contretemps goes unexplained in the records. Perhaps it was handled orally with the Liaison Officer.

However, on October 24, 1947, David L. Bazelon, assistant attorney general and director of the Office of Alien Property, wrote to the commanding officer of the Air Technical Service Command, Wright Field, Dayton, Ohio, regarding the Tesla photostats that had been sent by registered mail on or about September 11, 1945, to Colonel Holliday, at the latter's request.

"Our records do not reveal that this material has been returned," said Bazelon. He sent a description and asked that it be returned.

Obviously at least one set of Tesla's papers had reached Wright Field because on November 25, 1947, there was a response to the Office of Alien Property from Colonel Duffy, chief of the Electronic Plans Section, Electronic Subdivision, Engineering Division, Air Matériel Command, Wright Field. He replied: "These reports are now in the possession of the Electronic Subdivision and are being evaluated." He believed that the evaluation should be completed by January 1, 1948, and "At that time your office will be contacted with respect to final disposition of these papers."

There is no written record that OAP ever sought further to have the documents returned, and they were not returned.

For many years there have been rumors that these unpatented inventions or concepts of Tesla's found their way not only to the U.S. Army Air Force but to Russia and to private American defense industries, and ultimately into certain university research laboratories engaged in beam weaponry.

The Office of Alien Property experienced a very difficult problem over the years in explaining its role in connection with Tesla's papers. Between 1948 and 1978 it issued the following variations on a theme to many inquirers:

"While this Office participated in an examination of certain material owned by the late Dr. Tesla, our records do not disclose that any such material has been vested or is presently under the jurisdiction of this Office. . . ."

"This Office has never had custody . . . of any property of Nikola Tesla. . . ."

"While the Tesla papers were in our custody . . ."

"Photostatic copies of certain documents, made while the papers were under our seal. . . ."

"In 1943 this Office placed a seal on the property. . . ."

"While the Tesla papers were in our custody . . ." etc., etc., etc.

As for what is now Headquarters Aeronautical Systems Division, Wright-Patterson Air Force Base, Ohio, they state: "The organization (Equipment Laboratory) that performed the evaluation of Tesla's papers was deactivated several years ago. After conducting an extensive search of lists of records retired by that organization, *in which we found no mention of Tesla's papers, we concluded the documents were destroyed* at the time the laboratory was deactivated."[5] (Italics supplied. Response, under the Freedom of Information Act, dated July 30, 1980.)

Tesla's original papers, and the remaining models of his inventions—his magnifying transmitter, robot boats, early tube lighting, induction motors, turbine, exhibits shown at the Chicago World's Fair of 1893, such as the "Egg of Columbus," and others—left America in 1952 for Yugoslavia. His ashes were sent later. The artifacts may now be seen at the Tesla Museum in Belgrade, a dignified-looking building with a broad, well-proportioned facade at No. 51 Proleterskih Brigada, an avenue renamed after the war, but formerly known under the monarchy as Crown Street. The museum bears a plaque on a low wall, printed in the old Cyrillic alphabet.

Here Tesla's English writings have been translated into Serbo-Croatian—except, as the archivist admits, for the "unimportant" material, which remains, just as he wrote it, in the language of his adopted country.

30. THE LEGACY

The fact that Tesla's research notes and papers have not been easily available for western scientists has not, of course, meant that Teslian research is dead. On the contrary, the very mystery surrounding some of his unproved claims has served to goad numerous scientists into trying to duplicate his experiments. And since his aspirations were virtually limitless, there has always been a chance that the rewards of success would not be inconsiderable. But the single greatest stimulus to try to follow in Tesla's footsteps doubtless remains the example of the man himself—his stunning record of achievement and the enduring fascination of his mind. As one admiring German writer put it, "Tesla went beyond the borders of his exact science to foretell what lies in the future . . . a modern Prometheus who dared reach for the stars. . . ."[1]

Although a comprehensive summary of the state of Tesla-inspired research today would be beyond either the scope of this book or the intent of its author, no account of the inventor's life would be complete without at least some indication of what has become of a few of his major preoccupations. The record, as one might expect, is both mixed and incomplete, but it is no less impressive for that.

To begin, then, with Tesla's experiments with ball lightning: He had no idea what ball lightning might be useful for when he first encountered it in his Colorado Springs research; to him it was a nuisance, but it demanded an explanation. And so he set about determining the mode of formation of the strange fireballs and learned to produce them artificially.

The technical explanation runs like this: In the highly resonant transformer secondary comprising his magnifying transmitter, the entire energy accumulated in the excited circuit, instead of requiring a quarter period for transformation from static to kinetic, could spend itself in less time, at hundreds of thousands of horsepower. Thus, for example, Tesla

produced artificial fireballs by suddenly causing the impressed oscillations to be more rapid than free ones of the secondary. This shifted the
point of maximum electrical pressure below the elevated terminal capacity, and a ball of fire would leap great distances.

Yet strangely enough, modern plasma physicists with the best-
equipped laboratories, have failed to produce plasmoids with anything
near the stability of the true ball-lightning spheres that he created.

Why the fascination with this problem? First, of course, because
it is there, an unknown. But second because among other uses, it may
hold a vital key in the international race to achieve controlled nuclear
fusion—potentially the greatest power source in history. Among
those long interested in ball-lightning research have been Peter
Kapitza, the great Russian physicist, Lambert Dolphin and his colleagues in the radio physics laboratory at SRI International, Dr.
Robert W. Bass of Brigham Young University, and Robert Golka,
with whom Bass has collaborated on research.

Golka, a Massachusetts physicist, Tesla disciple, and lightning experimenter, has pursued the ephemeral fireball with the fervor of a hunter of
snarks. Like Tesla in Colorado, he has done his research alone in a remote
western laboratory in the Utah salt flats, and like Tesla, he has struggled
to win the kind of federal support that usually goes only to enormous
institutions or corporations.

In the largest hangar at the far end of the ghost base at Wendover,
Utah, which was built by the U.S. Army Air Force during World War II,
big spotlights are often burning as Golka conducts lightning tests. Here,
under tightest security in the 1940s, the B-29 *Enola Gay* was housed and
outfitted for delivering the first atomic bombs to Hiroshima.

Golka made two trips to the Tesla Museum to pore over the inventor's then unpublished notes and concentrated on replicating as exactly
as he could in the old air base hangar the magnifying transmitter that
Tesla had built in 1899 when investigating the lightning storms of Pike's
Peak.

"He [Tesla] was 'way ahead of anything we have today in the equipment he built," Golka says. "Such as the high-powered switches and spark gap switches. The knowledge has been lost; we don't know how he did it. Some of it was in the diaries, but he kept much of this stuff in his head."

Golka built a magnifying transmitter at his "Project Tesla" that would discharge 22 million volts, creating almost twice as powerful a chain-lightning storm as the maestro himself had produced at Colorado Springs.

The relevance of ball lightning to fusion research has to do with the problem of confining plasma. The heart of the most common type of experimental fusion reaction involves taking isotopic hydrogen gas and both accelerating and superheating it until the hydrogen nuclei fuse to make helium nuclei, releasing, in the process, staggering amounts of energy. Along the way, while the hydrogen is being charged with vast amounts of kinetic and thermal energy, it enters an imperfectly under-stood material state known as plasma.* In the penultimate stages of the process, before fusion begins, the besetting problem is to maintain the plasma's coherence, to confine it within some kind of invisible electro-magnetic "bottle."†

Since the strongest geometric shape is a sphere, Golka believes that ball lightning offers the best potential for containment of the unstable mass. He describes the odd lightning as "a glowing sphere of a variety of colors, a half-inch in diameter or as big as a grapefruit," and resembling an onion in its "layers and layers of alternate charged particles, positive and negative." It may bounce along through buildings, fall into water and set it boiling; and sometimes, as at the Hill Air Force Base in Utah, it

---

* Until recent years plasma had no major industrial importance but was merely a labo-ratory curiosity. Richard L. Bersin, executive vice president of International Plasma Corp., believes that the first practical application of plasmas came in the 19th century when "the glowing plasma produced by a Tesla coil was used to locate leaks in glass vacuum flasks."

† Teslian ideas are also involved in other aspects of fusion research. Superconducting magnetic coils, cooled to a few degrees above absolute zero, are used in magnetic con-tainment devices; and, in a rival process, hydrogen fuel pellets are being bombarded by high-energy particle beams.

may knock out the most sophisticated electronic equipment. In the summer of 1978, with the use of CO_2 laser beams, he finally managed to produce "bead" lightning, which he believes to be a form of ball lightning, and to photograph it in sequential frames.[2]

He then sought support from the U.S. Department of Energy for a major program of research for which he proposed to use a device called a pyrosphere, employing five laser beams to create thermonuclear fusion. In a "Fireball Fusion Reactor" only nonradioactive helium is created and, according to Golka, mathematical models indicate it can reach and hold temperatures above a billion degrees.

He also proposed to the Air Force another Teslian concept, a charged particle beam, but again one designed to employ laser technology. Such beam guns, he believes, would have a range of 6,000 miles and could melt and destroy ICBM-type missiles in the air. With a Tesla coil three times the size of his combined coils, Golka believed he could generate 200 million volts of electricity.

But he inherited the usual Teslian problems of a loner, and as he said, "The walls fall in on me when I work for corporations." His work reached a point where it could no longer progress with improvised equipment, but called for enormous investments. His competitors were large corporations and leading universities engaged in the nuclear-fusion race; and even some of the latter were being cut off from their federal grants. They too were deeply into laser technology, although Golka claims his system is different and unique. By no means the only scientist to have attempted to carry forward Tesla's work with ball lightning, he undoubtedly has been one of the most singleminded.*

Russia's Kapitza, who shared the 1978 Nobel Prize in physics with Arno Penzias and Robert W. Wilson of America for his work in magnetism and the behavior of matter at extremely low temperatures, acknowledges his debt to Tesla. "The efficient generation of super-high-frequency

* Labmert Dolphin says of Golka's replica of the Colorado Springs Tesla coil: "It is spectacular indeed, to either scientist or layman. I hope it ends up in a museum such as the Smithsonian where it can be appreciated." He too is a proponent of further research in ball lightning.

oscillations and their conversion back to direct-current electrical energy," he writes, "discloses possible solutions to the problem of transmitting electrical energy . . . in free space. The transmission setup will, of course, be similar to that already considered but, instead of a wave guide, a highly directional beam must be used, which, as is well-known, only at short wavelengths will diverge little. Such a setup for the transmission of electrical energy, firstly thought by N. Tesla many years ago, has already been discussed. . . . Although . . . possible in principle, it is tied up with the solution of a series of complicated engineering problems and therefore it can be implemented in practice only in such special situations in which other methods of energy transmission are inapplicable (for example, when energy must be supplied to a satellite)."[3]

In this field of wireless energy transmission, so directly concerned with the space race, there is progress nearer home. Richard Dickinson, who heads the Microwave Power Transmission project for Cal Tech's Jet Propulsion Laboratory in the desert near Barstow, California, traces his inspiration to the early work of Tesla. The concept of bringing electricity to Earth from an orbiting solar-power system via microwaves is daring, costly, romantic, and thoroughly in the style of the maestro.

"We beamed power from our transmitter at Goldstone a distance of one mile," Dickinson said of the NASA project initiated in the mid-seventies. "All of the microwave energy that fell within our target (of which we could only collect a portion with our existing apparatus), we converted 82.5 percent to useful direct current. Thirty-four thousand watts of direct current output carried a distance of one mile. We are well pleased. The next step is to look further into the technology and needs of the satellite power system of the future."[4]

William C. Brown of the Raytheon Company, who developed the rectenna used in this microwave-power research, also attributes the idea of sending electricity by radio waves to Tesla's pioneering in the fundamentals of radio broadcasting and wireless power transmission.

Theoretically, a city the size of New York could be supplied with five billion watts on a winter day by enormous satellite structures in the sky that would orbit synchronously with Earth at a height of 22,300 miles. But admittedly, the cost of such floating power stations would be many billions of dollars, and they would be highly vulnerable to enemy killer satellites in the event of war.

Brookhaven National Laboratory, located just to the northeast of Tesla's old Wardenclyffe site at Shoreham, also feels a close link with the inventor through the advanced high-energy work being conducted at the laboratory. In 1976 it paid homage to him in a ceremony, and the Yugoslav government sent a plaque to be placed at the still-standing Wardenclyffe laboratory.

Canada, too, has long been a bastion of Tesla Energy System advocates, and because of the country's rich hydroelectric sources, through-the-Earth transmission—if it worked—could be a boon to areas of power shortage.

But—will it work? Several projects have been planned, and some partially implemented, in Canada, central Minnesota, and most recently in Southern California—to "pump" hydroelectric power wirelessly through the Earth to an area of need, employing the Tesla system as it is understood.[5] The U.S. Department of Energy has often been asked to fund projects based on Tesla's system.

Unfortunately, there is no evidence that the system ever worked for Tesla, and none that it could work for anyone else. One of the inventor's problems was that he improperly extended into the electromagnetic domain fluid and fluid-mechanical analogies. Tesla's patent No. 787,412 provides for the Earth to be excited by a carefully valued wavelength to establish a standing wave condition. Tesla believed the propagation path fell along a diameter. But according to much knowledge developed since 1899, the propagation path would not be along a diameter but, rather, along an ellipsoidal arc somewhere between the diameter and the spherical surface.

A fundamental aspect of wave propagation of power is that *no power is transmitted if the wave is standing;* power is transmitted solely with a traveling component. Boundary layer propagation, i.e., the mode of lossless propagation of waves at the boundary of two differing media (such as earth and sky), is a viable concept. However, the boundary plane must be smooth and the waves must be properly launched. At the frequencies Tesla was using, such launching apparatus would be an enormous structure. In examining the photographs of his experimental station at Colorado Springs, it is apparent to experts that he did not employ apparatus essential to the launching of such waves.

Tesla probably was mistaken at Colorado Springs in his interpretation of the lightning storms which he observed traveling away from him (eastwardly) across the plains, producing maxima and minima effects upon his instruments. This he interpreted as standing waves being set up in the Earth by the traveling storm, with the crests of the waves passing through his location as the storms advanced. It is believed he was seeing an interference effect caused by the reradiating surface of the frontal range of mountains to the east of his station. The results would have been the same on his instruments.

Dr. Wait, formerly senior scientist at the Environmental Research Laboratories, National Oceanic and Atmospheric Administration, in Colorado, describes himself as a "firm skeptic" of the Tesla theory. "The concept that electromagnetic energy penetrates 'through the earth,'" he says, "is valid only if the frequency is sufficiently low *and* if the distances are small. It's all tied up with 'skin-effect' phenomena; that means that the field is confined to the surface of a good conductor as in metallic wave guide."[6]

Dr. Wait even goes so far as to suggest that Tesla never really accepted the fact that electromagnetic waves could transport energy through the air. "Instead he thought of the earth itself as a conveyor and also thought of the possibility of a return conductor at heights of '15 miles above sea level.' The parallel of this idea to the earth-ionosphere wave guide at extremely low frequencies is striking (see IEEE *Journals of Oceanic Engineering*, Vol. OE-2, No. 2, April 1977). Also his proposed resonance of the system might be interpreted as the first disclosure of the earth-ionosphere cavity oscillations that have been associated from the early 1960s with W. O. Schumann, N. Christofilos, and J. Galejs, among others."[7]

With respect to wireless communication, the U.S. Navy's Project Sanguine/Seafarer of recent years has evolved from Tesla's Colorado experiments. In a thermonuclear war, conventional radio communication probably would be disrupted at certain heights and wavelengths. America's atomic submarine fleet might then be without a means of receiving messages. The U.S. Navy, seeing this danger, turned back to Tesla's nineteenth-century suggestion of employing 10 Hz signals (ELF or extra-low frequency), to circle the globe and penetrate the deepest waters.

One of the headier speculations concerning Teslian science is a suggestion that Russia has been employing his theories on weather modification to interfere with the jet stream, causing droughts and extremes of hot and cold weather. However unlikely the charge, it is true that Tesla did do a good deal of theorizing (but very little experimentation) on weather control.

He wrote, for example, on the possible use of radio-controlled missiles and explosives to break up tornadoes and the use of "lightning of a certain kind" to trigger rainfall. Of the former he said, "It would not be difficult to provide special automata for this purpose, carrying explosive charges, liquid air or other gas, which could be put into action, automatically or otherwise, and which would create a sudden pressure or suction, breaking up the whirl. The missiles themselves might be made of material capable of spontaneous ignition." His proposal included a lengthy mathematical formula.[8]

As with much modern scientific exploration inspired by the maestro, the returns are still not in on weather changing. Scientist Frederic Jueneman, "Innovative Notebook" columnist for *Industrial Research* magazine, calls attention to the fact that Dr. Robert Helliwell and John Katsufrakis of Stanford University's Radio Science Laboratory, demonstrated that very low frequency radio waves can cause oscillations in the magnetosphere. With a 20-km antenna and a 5 kHz transmitter in the

Antarctic, they found that the earth's magnetosphere could be modulated to cause high-energy particles to cascade into our atmosphere, and by turning the signal on or off they could start or stop the energy flow.

"The theoretical implication suggested by their work," says Jueneman, "is that global weather control can be attained by the injection of relatively small 'signals' into the Van Allen belts—something like a super-transistor effect."

But Jueneman's speculations go further and are eminently worthy of Tesla: "If Tesla's resonance effects, as shown by the Stanford team, can control enormous energies by miniscule triggering signals, then by an extension of this principle we should be able to affect the field environment of the very stars in the sky. . . . With godlike arrogance, we someday may yet direct the stars in their courses."[9]

No biography of Tesla would be complete without mention of his bright following of amateur physicists who build Tesla coils for their personal research, endeavoring to replicate his electrical magic, and the young inventors who pore over his basic patents and still find inspiration from them.

Durlin C. Cox, a Wisconsin physicist who has pondered Tesla's published writings, has built two Tesla coils, the second of 10 million volts. The reasons: "My own personal interest in high voltage engineering, especially in the field of high frequency rf transformers; to further my studies on the laboratory production of ball lightning; and because the University of Wisconsin at Madison asked me to submit a Tesla coil in their bi-annual Engineering Exposition in the spring of 1981." He and friends built one Tesla coil for a Hollywood studio for lightning effects, which has been a common use of them.

Electrical engineer Leland Anderson has summarized the major points in design that a coil builder might gain from reading Tesla's *Colorado Springs Notes:*

1. The Q's of the primary and secondary must be as high as practicable.
2. The Q's of the primary and secondary should be equal.

3.   The length of the secondary winding should be one-quarter
of the effective operating wavelength.
4.   The technique of using an "extra coil" tank circuit (or a
variation of it) in the secondary to magnify the voltage should
be used.

"With these criteria in mind," he says, "the builder will find that
hundreds of turns are not necessary for the secondary winding to achieve
high voltages."

Last but not least, what about Tesla's death/disintegrator rays? Were his
concepts sound? If they were found useful by the U.S. Army Air Force
research team, whose top-secret project was rumored to have had the
code name "Project Nick," it may be safely assumed that instead of being
"destroyed," as reported, his papers are still highly classified.

Dr. Trump's evaluation and Swezey's assessment of Tesla's "secret
weapons" have, however, received updated concurrence by Lambert
Dolphin, assistant director of the Radio Physics Laboratory at SRI
International, who has studied the inventor's work and his ball-lightning
research for two decades. He points out that the fields of knowledge of
both physics and electrical engineering have grown exponentially since
about 1930.

"Whole libraries are now required just to keep track of all the theory
and experience that have unfolded since Tesla's time," he says. "Our
mathematical and practical understanding of electricity, magnetism, electromagnetic theory, and radio communications has continued to grow
explosively ever since 1950, or should I say 1970!"

Tesla, Dolphin believes, "may have had intuitive insight into lasers
and high-energy particle beams as well as ultra-high voltage phenomena,
but now that we understand all the physics much more, we can easily
evaluate many of his extravagant later-life claims."[10]

In fact, there is no good evidence to suggest that Tesla anticipated
lasers. His "teleforce rays" seem to have been concerned exclusively with
high-energy particle beams. We still do not know precisely how he

intended them to work, although, says Dolphin, the available evidence suggests that Tesla may not have paid sufficient attention to how greatly such beams may be absorbed or dispersed by molecules and atoms in the air. In any case, even if we did understand Tesla's intentions more clearly, we should be hard put to compare them to the current state of the art, much of which is hidden under high security classifications.

Nevertheless, Tesla's work with high voltages to accelerate charged particles does seem to have been decidedly in what is now the mainstream of physical research. "In this field," says Dolphin, "he anticipated modern linear and circular nuclear accelerators. Such machines today have energy levels of tens of billions of electron volts or at least 1,000 times greater energy levels than Tesla ever attained.

"I am sure his magnifying transmitters were spectacular. . . . He probably generated some interesting arcs and sparks that were what we now study as *plasmas*. The containment of plasmas is a huge area of modern physics. For example . . . to see if small amounts of matter can be turned into immense amounts of electrical power in carefully contained plasmas." But Tesla's early discoveries and inventions, he concludes, were indeed ingenious and ahead of their time.[11]

As this book goes to press, the Pentagon is studying the creation of a new branch of the armed services, to be known as the U.S. Space Command, whose primary arsenal will consist of laser and particle-beam weapons fired from "space battleships." In prose not unlike Tesla's own, a Department of Defense fact sheet compares particle beams to "directed lightning bolts"—although without explicitly admitting that such a weapon has in fact been developed.

It is difficult to assess the current state of the beam-weapons program because virtually everything about it is heavily classified. Apparently the technology involved has proved to be complex and difficult, raising questions about the project's feasibility, but many experts nevertheless seem to be hard at work on the problem. At the same time, the activities of the other nations in this area have been monitored carefully by agencies of the federal government. Indeed the possibility of creating a family of particle-beam weapons has been a subject of serious discussion in this country for at least the past twenty-five years, and it is, in my opinion, of no little significance that as long ago as 1947 the Military Intelligence Service identified the writings about a particle beam among Tesla's scientific papers as being "of extreme importance."

Since he had no laboratory in the later years of his life, Tesla was unable to develop his ideas. But it is undeniable that he described in general terms half a century ago what may prove to be one of the main weapons of the Space Age. And to the end of his days, Tesla the pacifist hoped that such knowledge would be used, not for war among Earthlings, but for interplanetary communication with our neighbors in space, of whose existence he felt certain.

BIBLIOGRAPHICAL ESSAY

Some of Tesla's own writing—lectures, articles, patents, papers, and letters—is available in the United States. His most important lectures and his brief autobiography, in bound volumes, are listed in the prologue to the reference notes.

Citation of biographies of Tesla by O'Neill, Hunt, and Draper, and others may be found in the reference notes. The O'Neill manuscript and the Swezey Collection are to be found at the Smithsonian Institution, Dibner Library.

Serious Tesla scholars will wish to consult the annotated *Dr. Nikola Tesla Bibliography* by J. T. Ratzlaff and L. I. Anderson (San Carlos, California, Ragusan Press, 1979), for it contains some 3,000 sources of writings by and about Tesla. "Priority in the Invention of Radio, Tesla v. Marconi" by Leland Anderson may be obtained through the Antique Wireless Association, Monograph New Series No. 4.

A new means of analyzing Tesla's inventions is provided in *Dr. Nikola Tesla: Selected Patent Wrappers from the National Archives* by J. T. Ratzlaff (Millbrae, Ca., Tesla Book Co., 1980). These "file wrappers" provide explanations and correspondence between the patentee and the Patent Office to overcome objections raised by the examiner.

Tesla's *Colorado Springs Notes, 1899–1900*, published in 1978 by the Tesla Museum, is available through the Tesla Book Company of Millbrae, California.

The Library of Congress Manuscripts Division contains microfilm correspondence between Tesla and George Scherff, Robert Underwood Johnson, Mark Twain, members of the Morgan family, George Westinghouse, and the Westinghouse Electric and Manufacturing Company.

In addition original correspondence and photographs may be found at the Butler Library, Rare Books and Manuscripts, Columbia University, including letters between Tesla and Johnson, Scherff, and others. The New York Public Library and the Engineering Societies Library, New York, have additional materials—the latter a large collection on legal proceedings for infringement of Tesla's AC patents.

Insight into the heyday of American invention and scientific and industrial growth are available in many publications but perhaps most colorfully in Matthew Josephson's *Edison* (New York, McGraw-Hill Book Co., 1959) and *The Robber Barons* (New York, Harcourt, Brace & World, Inc., 1934, 1962), Ronald W. Clark's *Edison* (New York, G. P. Putnam's Sons, 1977), and Robert A. Conot's *A Streak of Luck* (New York, Seaview Books, 1979).

See also Robert Silverberg's *Edison and the Power Industry*, Princeton, N.J., D. Van Nostrand Co., Inc., 1967), *The Electric Century 1874–1974*, reprint from *Electrical World*, McGraw-Hill, 1973; "Edisonian Vignettes," IEEE Spectrum, Vol. 15, No. 9 (September 1978); Francis Jehl, "Menlo Park Reminiscences," The Edison Institute, Dearborn, Mich., 1939, Vol. II, pp. 839–40; Alfred O. Tate, *Edison's Open Door* (New York, E. P. Dutton & Co., Inc., 1938); Daniel J. Kevles, *The Physicists* (New York, Alfred A. Knopf, 1978); W. A. Swanberg, *Citizen Hearst* (New York, Charles Scribner's Sons, 1961); Bernard Baruch, *Baruch, My Own Story* (New York, Henry Holt & Co., 1957); Margaret L. Coit, *Mr. Baruch* (Boston, Houghton Mifflin Co., 1957); and Henry G. Prout, *A Life of George Westinghouse* (New York, Charles Scribner's Sons, 1922, 1971).

Rebecca West's *Black Lamb, Gray Falcon* (New York, Viking, 1940, 1941) is a Westerner's first-person classic among many works attempting to sort out the complex pre–World War II history of the Yugoslavs. For an intimate recent view of an immigrant's life, read *Slovene Immigrant History* by Ivan Molek, translated from the manuscript *Over Hill and Dale* by Mary Molek (M. Molek, Inc., P.O. Box 453, Dover, Del., 19901). On the Communist revolution, *Memoir of a Revolutionary* by Milovan Djilas (New York, Harcourt, Brace, Jovanovich, 1973).

REFERENCE NOTES

Tesla's lectures and his own writings are to be found in the following: Nikola Tesla, *Lectures, Patents, Articles*, Nikola Tesla Museum, 1956, reprinted 1973 by Health Research, Mokelumne Hills, California 95245; lectures in part in Thomas Commerford Martin's *Inventions, Researches and Writings of Nicola Tesla*, originally published in 1894 in *The Electrical Engineer*, New York, and republished 1977 by Omni Publications, Hawthorne, California 90250. See also: "My Inventions," Tesla's autobiography (which appeared originally in the *Electrical Experimenter*, May, June, July, October, 1919), republished by Školska Knjiga, Zagreb, Yugoslavia, 1977, with the Nikola Tesla Museum.

Included in the first two volumes are these important lectures: "A New System of Alternate Current Motors and Transformers," American Institute of Electrical Engineers, New York, May 16, 1888, describing his polyphase system of alternating current; "Experiments with Alternate Currents of Very High Frequency and Their Application to Methods of Artificial Illumination," AIEE, Columbia College, May 20, 1891; "Experiments with Alternate Currents of High Potential and High Frequency," IEE, London, February 3, 1892; repeated before Royal Institution, London, February 4; and again February 19 in Paris before the Société Internationale des Electriciens and the Société Française de Physique. In these he introduces the Tesla coil for high-frequency, high-voltage research effects.

They also include "On Light and Other High Frequency Phenomena," February 24, 1893, before the Franklin Institute in Philadelphia, and again in St. Louis before the National Electric Light Assn., March 1. Here he covered the principles of radio communication. "Mechanical and Electrical Oscillators," August 25, 1893, before the International

Electrical Congress at the World's Fair, Chicago. "On Electricity," at the
Ellicott Club, Buffalo, to commemorate Niagara Falls power, January
12, 1897. "On the Streams of Lenard and Röntgen, with Novel Appara-
tus for Their Production," before the New York Academy of Sciences,
April 6, 1897. "High Frequency Oscillators for Electro-therapeutic and
Other Purposes," before the Electrotherapeutic Association, Buffalo,
September 13, 1898.

Tesla's *Colorado Springs Notes, 1899–1900*, published in 1978 by
the Tesla Museum in Yugoslavia, is available through the Tesla Book
Company of Millbrae, California.

CHAPTER 1. *MODERN PROMETHEUS*

1. John J. O'Neill, *Prodigal Genius*, (New York, David McKay Co.,
   1944), pp. 93–95, 283; Inez Hunt and W. W. Draper, *Lightning in
   His Hand* (Hawthorne, Calif., Omni Publications, 1964, 1977),
   pp. 54–55.
2. Microfilm letters, Twain to Tesla, Library of Congress, n.d.
3. Chauncey McGovern, "The New Wizard of the West," *Pearson's
   Magazine*, London, May 1899.
4. O'Neill, *Genius*, p. 158.

CHAPTER 2. *A GAMBLING MAN*

1. Nikola Tesla, "My Inventions," *Electrical Experimenter*, May, June,
   July, October 1919, republished by Skolska Knjiga, Zagreb, Yugoslavia, 1977, p. 30.
2. Ibid, pp. 30–31.
3. Ibid., p. 26.
4. Ibid., pp. 8–9.
5. Ibid., p. 17.
6. Ibid, p. 18.
7. Ibid., p. 9–10.
8. Ibid., p. 10–12.
9. Ibid., p. 12–13.
10. Ibid., p. 12.
11. Ibid., p. 13.
12. Ibid., p. 13.
13. Ibid., p. 14.
14. Ibid., p. 16.
15. Ibid., p. 14.
16. Ibid., p. 35–36.

17. Ibid.
18. O'Neill, *Genius*, pp. 36–37.
19. Tesla, "Inventions," p. 41.
20. Nikola Trbojevich, *Spomenica* (Anniversary Booklet of the Serb National Federation), 1901–51, Pittsburgh, Pa., p. 172. Source: Immigrant Archives, University of Minnesota Library.
21. Tesla, "Inventions," p. 18.

CHAPTER 3. *IMMIGRANTS OF DISTINCTION*

1. Tesla, "Inventions," pp. 42–44.
2. Ibid., p. 43.
3. Ibid., p. 44.
4. Kenneth M. Swezey, "Nikola Tesla," *Science,* Vol. 127, No. 3307 (May 16, 1956), p. 1148. O'Neill, *Genius,* pp. 48–51.
5. Tesla, "Inventions," p. 46.
6. Ibid., p. 46.
7. Ibid., p. 48.
8. Ibid., p. 50.
9. Ibid., p. 50.

CHAPTER 4. *AT THE COURT OF MR. EDISON*

1. Matthew Josephson, *Edison* (New York, McGraw-Hill Book Co., 1959).
2. Ibid.
3. O'Neill, *Genius*, p. 60.
4. Tesla, "Inventions," p. 51.
5. Ibid., p. 54.
6. O'Neill, *Genius*, p. 64.
7. Josephson, *Edison*, pp. 87–88.
8. New York *Times,* October 19, 1931.
9. Josephson, *Edison*.
10. O'Neill, *Genius*, p. 64.
11. Matthew Josephson, *The Robber Barons* (New York, Harcourt, Brace & World, Inc., 1934, 1962).
12. Ibid.
13. O'Neill, *Genius*, p. 64; *Electrical Review,* New York, August 14, 1886, p. 12.
14. O'Neill, *Genius*, p. 66.

CHAPTER 5. *THE WAR OF THE CURRENTS BEGINS*

1. *Electrical Review,* May 12, 1888, p. 1; "Nikola Tesla," Swezey, p. 1149; O'Neill, *Genius*, pp. 67–68.

2. O'Neill, *Genius,* p. 69.
3. B. A. Behrend, Minutes, Annual Meeting American Institute of Electrical Engineers, New York, May 18, 1917, Smithsonian Institution.
4. Josephson, *Edison.*
5. Ibid., p. 346.
6. Ibid., p. 346.
7. Ibid., p. 349.
8. Ibid., p. 347.
9. Ibid., p. 349.
10. Josephson, *The Robber Barons.*
11. Ibid.
12. Ibid.
13. O'Neill, *Genius,* p. 84.
14. Ibid., p. 81.
15. Ibid., p. 82.
16. Speech, Institute of Immigrant Welfare, Hotel Biltmore, New York, May 12, 1938, read in absentia.
17. Letter to Tesla from Michael Pupin, December 19, 1891, Tesla Museum, Belgrade.
18. Hunt and Draper, *Lightning.*

CHAPTER 6. *ORDER OF THE FLAMING SWORD*

1. "Experiments with Alternate Currents of Very High Frequency," a lecture at Columbia College by Tesla on May 20, 1891.
2. T. C. Martin, ed.: *The Inventions, Researches and Writings of Nikola Tesla* (Hawthorne, California, Omni Publications, 1977), pp. 200–201.
3. Ibid., p. 236.
4. Ibid., pp. 245–64; also O'Neill, *Genius,* pp. 150–54.
5. O'Neill, *Genius,* pp. 146–49.
6. Ibid., pp. 152–53.
7. Ibid., pp. 150–51. See also Tesla's lecture of February 1892 before the Royal Society of Great Britain and the Society of Electrical Engineers of France, Paris.
8. Martin, *Inventions,* p. 261.
9. *The Story of Science in America* (New York, Charles Scribner's Sons, 1967).
10. Testimonial from Maj. Edwin H. Armstrong on Tesla's seventy-fifth birthday, Tesla Museum, Belgrade, Yugoslavia, n.d.
11. Letter to Tesla from J. A. Fleming, 1892, Tesla Museum, Belgrade.
12. O'Neill, *Genius,* p. 88.
13. Nikola Tesla, "Massage with Currents of High Frequency," *Electrical Engineer,* December 23, 1891, p. 697; Martin, *Inventions,* p. 394; O'Neill, *Genius,* p. 91; Nikola Tesla, Lectures, Patents, Arti-

cles, Nikola Tesla Museum, 1956; reprinted 1973 by Health Research, Mokelumne Hills, California 95245, p. L-156, Lecture to American Electro-Therapeutic Assn., Buffalo, September 13, 1898.

CHAPTER 7. *RADIO*

1. Tesla, "Inventions," p. 69.
2. Ibid., p. 62.
3. Letter from Sir William Crookes to Tesla, March 8, 1892, Tesla Museum, Belgrade, Yugoslavia.
4. Tesla, "Inventions," p. 80.
5. Ibid., p. 81.
6. Ibid., p. 82.
7. O'Neill, *Genius,* p. 264.
8. Tesla, "Inventions," p. 62.
9. O'Neill, *Genius,* pp. 131–34; *United States Reports,* Cases Adjudged in the Supreme Court of the United States, Vol. 320, Oct. Term, 1942: *Marconi Wireless Telegraph Company of America* v. *United States,* pp. 1–80; L. I. Anderson, "Priority in Invention of Radio, *Tesla* v. *Marconi,*" Antique Wireless Assn., March 1980, monograph; see also abbreviated translation, *Voice of Canadian Serbs,* Chicago, July 16, 1980.
10. *United States Reports,* "Transcript of Record," p. 979. Also: Anderson, "Priority."
11. Martin, *Inventions,* pp. 477–85.
12. Paper by Tesla for birthday press conference, around 1938. See also lecture by Tesla to American Electro-Therapeutic Assn., Buffalo, N.Y., Sept. 13, 15, 1898.
13. Martin, *Inventions,* pp. 486–93.

CHAPTER 8. *HIGH SOCIETY*

1. Bernard Baruch, *My Own Story* (New York, Henry Holt & Co., 1957).
2. Julian Hawthorne Papers, Bancroft Library, University of California, Berkeley.
3. Arthur Brisbane, "Our Foremost Electrician," New York *World,* July 22, 1894, p. 17. Also *Electrical World,* August 4, 1894, p. 27.
4. O'Neill, *Genius.*
5. O'Neill, *Genius,* pp. 288–89.
6. Julian Hawthorne Papers.
7. Waldemar Kaempffert, "Electrical Sorcerer," New York *Times* Book Reviews, February 4, 1945, pp. 6, 22.
8. O'Neill, *Genius,* p. 167.
9. Margaret Storm, *Return of the Dove* (Baltimore, Maryland, Margaret Storm Publication, 1959).

10. Tesla, "Inventions," p. 78.
11. Swezey, "Nikola Tesla," p. 1158.
12. Hunt and Draper, *Lightning*, p. 199.
13. O'Neill, *Genius*, pp. 302–03.
14. Ibid., p. 303.

CHAPTER 9. *HIGH ROAD, LOW ROAD*
1. Swezey, "Nikola Tesla," p. 2. Also, O'Neill, *Genius*, pp. 103–06.
2. Swezey, "Nikola Tesla," p. 3.
3. B. A. Behrend, "Dynamo-Electric Machinery and Its Evolution," *Western Electrician*, September 1907.
4. Tesla, "Inventions," p. 63.
5. O'Neill, *Genius*, pp. 238–43.
6. Martin, *Inventions*, p. 292.
7. Letter, Katharine Johnson to Tesla, February 1894, Tesla Museum, Belgrade, Yugoslavia.
8. Letter, Tesla to Katharine Johnson, May 11, 1894, Rare Books & Manuscripts, Butler Library, Columbia University.
9. Letter, Katharine Johnson to Tesla, June 15, 1894, Tesla Museum, Belgrade, Yugoslavia.
10. New York *Times*, March 14, 1895, p. 9; New York *Herald*, March 14, 1895, *Electrical Review*, March 20, 1895.
11. *Electrical Review*, London, March 15, 1895, p. 329.
12. Charles Dana, New York *Sun*, March 14, 1895, p. 6 (editorial).
13. Letter, Katharine Johnson to Tesla, March 14, 1895, Tesla Museum, Belgrade.

CHAPTER 10. *AN ERROR OF JUDGMENT*
1. Microfilm letter, Tesla to Alfred Schmid, March 30, 1895, Library of Congress.
2. Microfilm letter, Tesla to Alfred Schmid, April 3, 1895, Library of Congress.
3. Michael Pupin, *From Immigrant to Inventor* (New York, Charles Scribner's Sons, 1922).
4. Ibid.; see also Josephson, *Edison*, pp. 381–83; Nikola Tesla, "On Roentgen Rays," *Electrical Review*, March 11, 1896, pp. 131–35. Same issue, "Tesla Radiographs," p. 134; also March 18, pp. 146, 147; April 8, 1896, pp. 180, 183, 186. See also *Electrical World*, March 28, 1896, 343–44.
5. Pupin, *Immigrant*.
6. *Electrical Review*, New York, April 14, 1897, p. 175; see also Nikola Tesla, *Colorado Springs Notes, 1899–1900:* commentaries, Aleksandar Marinčić, p. 398, Tesla Museum, Belgrade.

7. Letter, Prof. Walter Thumm, Queen's University, Ontario, Canada, to Nick Basura, May 23, 1975, p. 2.
8. *Electrical Review*, March 11, 18, April 8, 1896.
9. Josephson, *Edison*, p. 382.
10. Robert Conot, *Streak of Luck* (New York, Seaview Books, 1979).
11. Letter, Dr. Lauriston S. Taylor to author, 1980.
12. Tesla, "On Roentgen Rays," p. A-31.
13. New York *Times*, March 12, 1896, p. 9, col. 3.
14. Nikola Tesla, "Tesla on Hurtful Actions of Lenard and Roentgen Rays," *Electrical Review*, May 5, 1897, pp. 207–11.
15. New York *Herald*, undated anonymous article written two years after Tesla's laboratory fire of March 13, 1895. Butler Library, Columbia University.
16. Ibid.

CHAPTER 11. *TO MARS*

1. Letter, Katharine Johnson to Tesla, April 3, 1896, Tesla Museum, Belgrade, Yugoslavia.
2. Letter, Katharine Johnson to Tesla, summer 1896, Tesla Museum, Belgrade, Yugoslavia.
3. Letter, Robert U. Johnson to Tesla, January 10, 1896, Columbia University, Butler Library.
4. Microfilm letter, Robert U. Johnson to Tesla, October 25, 1895, Library of Congress.
5. Letter, Tesla to Robert U. Johnson, March 13, 1896, Butler Library, Columbia University.
6. Letter, Katharine Johnson to Tesla, December 26, 1896, Tesla Museum, Belgrade.
7. *Electrical Review*, August 11, 1897; see also New York *Sun*, August 4, 1897.
8. Anderson, "Priority."
9. *Electrical Review*, August 11, 1897, p. X. See also New York *Sun*, August 4, 1897.
10. *Electrical Engineer*, London, August 20, 1897, p. 225. From New York *Journal*, August 4, 1897, p. 1.
11. *Electrical Review*, August 11, 1897; *Electrical Engineer*, New York, June 23, 1897, p. 713.
12. *Electrical Review*, August 11, 1897.
13. *Electrical Review*, March 29, 1899, p. 197.
14. Ibid.
15. Letter, Katharine Johnson to Tesla, January 12, 1896, Tesla Museum, Belgrade.
16. O'Neill, *Genius*, pp. 161–62.
17. A. L. Benson, *The World Today*, Vol. XXI, No. 8 (February 1912).

CHAPTER 12. *ROBOTS*

1. *Mining & Scientific Press,* January 15, 1898, p. 60.
2. McGovern, "The New Wizard."
3. *Century* magazine, "Tesla's Oscillator and Other Inventions, p. 922, April 1895. See also *Electrical Review,* Volume 34, No. 13 (March 29, 1899).
4. New York *Times,* January 6, 1898, p. 5, col. 5.
5. *Electrical Review,* New York, January 5, 1898.
6. W. A. Swanberg, *Citizen Hearst* (New York, Charles Scribner's Sons, 1961).
7. Ibid.
8. Philadelphia *Press,* May 1, 1898.
9. Ibid.
10. Cdr. E. J. Quinby, USN Ret., "Communications: Encoded, Decoded, Codeless," *Dots and Dashes,* Vol. 5, No. 1, Lincoln, Neb., January, February, March 1976.
11. Swezey, "Nikola Tesla," pp. 1155–56.
12. O'Neill, *Genius,* 166–74.
13. Microfilm letter, Mark Twain to Tesla, November 17, 1898, Library of Congress.
14. Letter, Tesla to Katharine Johnson, November 3, 1898, Butler Library, Columbia University.
15. N. G. Worth, "An Inquiry About Tesla's Electrically Controlled Vessel," *Electrical Review,* New York, November 30, 1898, p. 343.
16. Microfilm letter, Tesla to Robert U. Johnson, December 1, 1898, Library of Congress.
17. "Science and Sensationalism," *Public Opinion,* December 1, 1898, pp. 684, 685.
18. Tesla, "Inventions," p. 84. See also *Electrical Experimenter,* January, February, March, April, May 1919.
19. Tesla, "Inventions," pp. 84, 85.
20. Ibid., p. 85.
21. Ibid., p. 85.
22. Letter, Tesla to B. F. Meissner, Manuscript Division, Library of Congress, September 29, 1915.
23. Letter, Leland Anderson to Nick Basura, March 4, 1977.
24. Ibid.
25. New York *Times,* February 1, 1944, editorial.
26. Letter, Tesla to Leonard Curtis, 1899.

CHAPTER 13. *HURLER OF LIGHTNING*

1. O'Neill, *Genius,* pp. 175–76.
2. Letter, Tesla to Katharine Johnson, March 9, 1899, Special Collections, Butler Library, Columbia University.

3. Letter, Tesla to Robert U. Johnson, March 25, 1899, Special Collections, Butler Library, Columbia University.
4. Tesla, "Inventions," pp. 64–67; *Electrical Experimenter,* June 1919, pp. 112–76.
5. Ibid.
6. Tesla, *Colorado Springs Notes,* pp. 127–33, 165, Tesla Museum with Nolit, Belgrade, Yugoslavia, 1978.
7. Tesla, *Colorado Springs Notes,* pp. 167, 168; Leland I. Anderson, "Wardenclyffe—A Forfeited Dream," *Long Island Forum,* August, September 1968. See also *The Teslian,* November 1955, Butler Library, Columbia University.
8. James R. Wait, "Propagation of ELF Electromagnetic Waves and Project Sanguine/Seafarer," *IEEE Journal of Oceanic Engineering,* Vol OE-2, No. 2 (April 1977).
9. Microfilm letter, George Scherff to Tesla, early 1899, Library of Congress.
10. Microfilm letter, Tesla to Scherff, April 13, 1899, Library of Congress.
11. Microfilm letter, Tesla to Robert U. Johnson, August 16, 1899, Library of Congress.
12. Tesla, *Colorado Springs Notes,* pp. 127–33.
13. Nikola Tesla, "Transmission of Energy Without Wires," *Scientific American Supplement,* June 4, 1904, pp. 23760–1. (Reprint of *Electrical World & Engineering,* March 5, 1904; description of Colorado Springs experiments.)
14. Ibid. See also O'Neill, *Genius,* pp. 179–81.

CHAPTER 14. *BLACKOUT AT COLORADO SPRINGS*

1. Tesla, "Transmission"; O'Neill, *Genius,* p. 180.
2. Tesla, "Transmission"; Tesla, *Colorado Springs Notes,* p. 62.
3. Ibid.
4. Ibid.
5. Tesla, *Colorado Springs Notes.* Microfilm letter, Tesla to George Westinghouse, January 22, 1900, Library of Congress.
6. O'Neill, *Genius,* 183–87.
7. Tesla, *Colorado Springs Notes,* p. 29.
8. Nikola Tesla, Minutes of the Edison Medal Meeting, American Institute of Electrical Engineers, May 18, 1917, Smithsonian Institution.
9. Ibid.
10. Microfilm letter, Tesla to Robert U. Johnson, August 16, 1899, Library of Congress.
11. Ibid.
12. O'Neill, *Genius,* p. 189.

13. Nikola Tesla, "The Problem of Increasing Human Energy," *Century* magazine, June 1900, p. 210.
14. Tesla, *Colorado Springs Notes,* pp. 368–70.
15. Ibid.
16. Nikola Tesla, "Talking With Planets," *Current Literature,* March 1901, p. 359; also Colorado Springs *Gazette,* March 9, 1901, p. 4, col. 2.
17. Colorado Springs *Gazette,* loc. cit.

CHAPTER 15. *MAGNIFICENT AND DOOMED*

1. Tesla, *Colorado Springs Notes,* p. 367.
2. Ibid., p. 370.
3. Ibid.
4. Microfilm letter, Tesla to George Westinghouse, January 22, 1900, Library of Congress.
5. Microfilm letter, Tesla to Robert U. Johnson, early 1900, Library of Congress.
6. Microfilm letter, Tesla to J. Pierpont Morgan, November 26, 1900, Library of Congress.
7. Microfilm letter, Tesla to Morgan, December 12, 1900, Library of Congress.
8. Microfilm letter, Morgan to Tesla, February 15, 1901, Library of Congress.
9. Anderson, "Wardenclyffe."
10. Ibid.
11. Microfilm letter, Tesla to Stanford White, September 13, 1901, Library of Congress.

CHAPTER 16. *RIDICULED, CONDEMNED, COMBATTED*

1. Seattle Sunday *Times,* Don Duncan's "Driftwood Days," July 1972.
2. Ibid.
3. Philadelphia *North American,* "Lord Kelvin Believes Mars Now Signalling America"; "Tesla Thinks Wind Power Should be Used More Now"; May 18, 1902, Mag. Sec. V.
4. Ibid.
5. Philadelphia *North American,* n.d., Julian Hawthorne Papers, Bancroft Library, University of California.
6. Ibid.
7. Letter, Tesla to Hawthorne, n.d., Julian Hawthorne Papers, Bancroft Library, University of California.
8. New York *Times,* "Court Excuses Tesla," October 16, 1902.

9. Microfilm letter, Tesla to George Scherff, n.d., Library of Congress.
10. Microfilm letter, Tesla to J. Pierpont Morgan, April 8, 1903, Library of Congress.
11. Microfilm letter, Tesla to Morgan, July 3, 1903, Library of Congress.
12. Microfilm letter, Morgan to Tesla, July 14, 1903, Library of Congress.
13. Letter, Richmond Pearson Hobson to Tesla, n.d. (referring to Hobson letter of May 6, 1902), Tesla Museum, Belgrade, Yugoslavia.
14. Letter, Tesla to George Scherff, July 18, 1905, Butler Library, Columbia University.
15. Conversation, L. Anderson and Dorothy F. Skerritt, March 24, 1955.
16. Letter, Katharine Johnson to Tesla, n.d., Special Collections, Butler Library, Columbia University.
17. Letter, Robert U. Johnson to Tesla, n.d., Special Collections, Butler Library, Columbia University.
18. Letter, R. P. Hobson to Tesla, May 1, 1905, Tesla Museum, Belgrade, Yugoslavia.
19. Letter, Katharine Johnson to Tesla, n.d., Tesla Museum.
20. Letter, Tesla to Katharine Johnson, n.d., Tesla Museum.
21. Letter, Katharine Johnson to Tesla, n.d., Tesla Museum.
22. Letter, Tesla to George Scherff, October 26, 1905, Special Collections, Butler Library, Columbia University.
23. Swezey, "Nikola Tesla."
24. Letter, Tesla to George Westinghouse, January 11, 1906, Special Collections, Butler Library, Columbia University.
25. Anderson, "Wardenclyffe."
26. Brooklyn *Eagle*, March 26, 1916.
27. Anderson, "Wardenclyffe."
28. *Electrical Experimenter*, "U.S. Blows Up Tesla Radio Tower," September 1917, p. 293; *Literary Digest*, "Spies & Wireless," September 1, 1917, p. 24.
29. Microfilm letter, Tesla to Scherff, July 13, 1913, Library of Congress.
30. Microfilm letter, Tesla to Morgan, July 13, 1913, Library of Congress.

CHAPTER 17. *THE GREAT RADIO CONTROVERSY*

1. Charles Süsskind, *Dictionary of American Biography*, Supp. 3 (New York, Charles Scribner's Sons, 1941–45), pp. 767–70.
2. Wait, "Propagation of ELF Electromagnetic Waves."
3. Anderson, "Priority."
4. Ibid.

5. Ibid.
6. Gen. T. O. Mauborgne, "Tesla the Wizard," *Radio-Electronics,* February 1943.
7. Brooklyn *Standard Union,* May 12, 1910.
8. Cdr. E. J. Quinby, letter to author, November 19, 1977. See also Quinby, "Nikola Tesla," *Proceedings, Radio Club of America,* Fall 1971.
9. Los Angeles *Examiner,* May 13, 1915, "Prof. Pupin Now Claims Wireless His Invention."
10. Ibid.
11. Letter, Armstrong to Anderson, November 16, 1953, Butler Library, Columbia University. See also Edwin Armstrong, "Progress of Science," *Scientific Monthly,* April 1943, pp. 378–81.
12. Letter, Anderson to author, November 5, 1977.
13. Haraden Pratt, "Nikola Tesla, 1856–1943," *Proceedings of the IRE,* September 1956.
14. Dragislav L. Petković, "A Visit to Nikola Tesla," *Politika,* Belgrade, April 27, 1927, No. 6824.

CHAPTER 18. *MIDSTREAM PERILS*

1. Joseph S. Ames, "Latest Triumph of Electrical Invention," *Review of Reviews,* June 1901.
2. F. P. Stockbridge, "The Tesla Turbine," *The World's Work,* March 1912, pp. 534–48. See also Nikola Tesla, "Tesla's New Method of and Apparatus for Fluid Propulsion," *Electrical Review & Western Electrician,* September 9, 1911, pp. 515–17; New York *Times,* "Tesla's New Engine," September 13, 1911. U.S. Patent Office: Patent 1,061,142, Fluid Propulsion, May 6, 1913; 1,061,206, Turbine, May 6, 1913; 1,329,559, Valvular Conduit, February 3, 1920.
3. Stockbridge, "Turbine."

CHAPTER 19. *THE NOBEL AFFAIR*

1. Microfilm letter, J. P. Morgan Company to Tesla, May 25, 1913, Library of Congress.
2. Microfilm letter, Tesla to J. P. Morgan, May 19, 1913, Library of Congress.
3. Microfilm letter, Tesla to J. P. Morgan, June 19, 1913, Library of Congress.
4. Microfilm letter, Robert U. Johnson to Tesla, April 22, 1913, Library of Congress.
5. Microfilm letter, Tesla to Robert U. Johnson, May 9, 1913, Library of Congress.
6. Cleveland Moffett, "Steered by Wireless," Tesla-Hammond corre-

spondence, 1910–1914, L. Anderson collection; *McClure's Magazine,* March 1914.

7. "The Goldschmidt Radio Tower," *Electrical Experimenter,* February 1914, p. 154. Same issue: H. Winfield Secor, "Currents of Ultra-High Frequency," pp. 151–54.

8. New York *Times,* "Edison and Tesla to Get Nobel Prizes," November 6, 1915, p. 1, col. 4. New York *Times,* November 7, 1915, II, p. 17, col. 3.

9. Ibid. New York *Times,* November 7, 1915.

10. New York *Times,* November 14, 1915.

11. Microfilm letter, Tesla to Robert U. Johnson, November 29, 1919, Library of Congress.

12. *Literary Digest,* "Three Nobel Prizes for Americans," December 1915, p. 1426.

13. "The Nobel Prize," *Electrical World,* November 13, 1915.

14. O'Neill, *Genius,* p. 229.

15. Hunt and Draper, *Lightning,* p. 170.

CHAPTER 20. *FLYING STOVE*

1. Microfilm letter, Tesla to Westinghouse Company, July 7, 1912, Library of Congress.

2. *"Teslin 'Ventilni Vod' I Fluidika,"* Prof. Tugomir Šurina, Symposium Nikola Tesla, Yugoslavia, 1976.

3. Warren Rice, "An Analytical & Experimental Investigation of Multiple Disc Pumps & Turbines," *Journal of Engineering for Power. Trans. ASME* Vol. 85, Series A, No. 3 (July 1963), Paper No. 62-WA-191, pp. 191–98; also Vol. 87, Series A, No. 1 (January 1965), Paper No. 63-WA-67, pp. 29–36. See also *ASME Transactions* of 1970s.

4. SunWind Ltd., Newsletter No. 10, March 12, 1979, Sebastopol, Calif. 95472.

5. Tesla letter to New York *Times,* September 15, 1908. See also New York *Herald Tribune,* July 12, 1927, "Tesla Predicts Fuelless Plane."

6. Tesla, New York *Times,* September 15, 1908.

7. New York *Times,* "Tesla Gets Patent on Helicopter-Plane," February 22, 1928, p. 18, col. 4. *Science & Invention,* June 1928, p. 116.

8. *Review, The Yugoslav Monthly Magazine,* July–August 1964, "Helicopter in Hansom Cab Days," pp. 31–33.

9. Microfilm letter, Tesla to Scherff, July 1, 1909, Library of Congress.

10. Microfilm letter, Tesla to Scherff, n.d., Library of Congress.

11. Microfilm letter, Tesla to Scherff, October 15, 1918, and Scherff to Tesla, October 1918, Library of Congress.

12. Letter, Tesla to Anne Morgan, March 31, 1913, Tesla Museum, Belgrade, Yugoslavia.
13. Letter, Anne Morgan to Tesla, May 3, 1913, Tesla Museum, Belgrade, Yugoslavia.
14. Letter, Tesla to Anne Morgan, May 7, 1913, Tesla Museum, Belgrade, Yugoslavia.
15. Letter, Anne Morgan to Tesla, April 26, 1926, Tesla Museum, Belgrade, Yugoslavia.
16. Letter, Katharine Johnson to Tesla, n.d., Tesla Museum, Belgrade, Yugoslavia.

CHAPTER 21. *RADAR*

1. New York *Times*, March 18, 1916, p. 8, col. 3.
2. Speech by Millikan at Chemists' Club, New York, October 7, 1928.
3. The Royal Bank of Canada, *Monthly Letter*, Vol. 59, No. 11 (November 1978).
4. New York *Times*, December 8, 1915, p. 8, col. 3, See New York *Herald*, April 15, 1917.
5. Dr. Emil Girardeau, "Nikola Tesla, Radar Pioneer," translation from the French, presented at Nikola Tesla—Kongress, Vienna, September 1953.
6. Ibid. See also, Nikola Tesla, "The Problem of Increasing Human Energy," *Century* magazine, June 1900, pp. 208–09; New York *Times*, "America's Invisible Airplane," September 7, 1980, p. 20 E.

CHAPTER 22. *THE GUEST OF HONOR*

1. O'Neill, *Genius*, p. 230.
2. Ibid., p. 231.
3. *Minutes, Edison Medal Meeting*, American Institute of Electrical Engineers, May 18, 1917, Smithsonian Institution.
4. Ibid.
5. Ibid.
6. Ibid.
7. Petković, "A Visit to Nikola Tesla."
8. Letter, Tesla to Scherff, March 3, 1918, Butler Library, Columbia University.
9. Letter, Scherff to Tesla, June 23, 1916, Butler Library, Columbia University.
10. Microfilm letter, Tesla to Scherff, October 15, 1918, Library of Congress.
11. Microfilm letter, Tesla to Robert U. Johnson, December 27, 1914, Library of Congress.
12. Microfilm letter, Robert U. Johnson to Tesla, December 30, 1919, Library of Congress.

13. Letter, Katharine Johnson to Tesla, n.d., Tesla Museum, Belgrade, Yugoslavia.

CHAPTER 23. *PIGEONS*

1. Microfilm letter, Tesla to E. M. Herr, president of Westinghouse, November 13, 1920, Library of Congress.
2. Microfilm letter, Westinghouse Electric Company to Tesla, November 28, 1921, Library of Congress.
3. Microfilm letter, Tesla to Westinghouse Electric Company, 1921, Library of Congress.
4. Microfilm letter, Tesla to Westinghouse, January 22, 1922, Library of Congress. Also February 23 and March 11, 1922.
5. Microfilm letter, Tripp to Tesla, early 1922, Library of Congress.
6. "Hundredth Anniversary Nikola Tesla—and Ivan Meštrović," *Enjednicar*, Serbian Cultural Society Education, Croatian Serbs, Zagreb, Yugoslavia, April 11, 1956, pp. 1–2.
7. "Secanja na Teslu Kenet Suizia" ("Kenneth Swezey's Recollections of Nikola Tesla"), *Tesla*—Belgrade, IV (1957), 38–39, pp. 45–48.
8. K. M. Swezey, "Nikola Tesla," *Psychology*, October 1927, p. 60.
9. Nikola Tesla, "A Story of Youth Told by Age," Smithsonian Institution.
10. O'Neill, *Genius*, pp. 309–10.
11. Ibid., pp. 311–12.
12. Ibid., pp. 315–17.
13. Jule Eisenbud, "Two Approaches to Spontaneous Case Material," *Journal of American Society for Psychic Research*, July 1963.
14. Ibid.
15. Ibid.

CHAPTER 24. *TRANSITIONS*

1. Detroit *Free Press*, August 10, 1924, Feature sec., p. 4. See also, *Collier's*, "When Woman Is Boss," January 30, 1926.
2. *Collier's*, op. cit.
3. Ibid.
4. Microfilm letter, Johnson to Tesla, April 9, 1925, Library of Congress.
5. Microfilm letter, Tesla to Johnson, June 3, 1925, Library of Congress.
6. Letter, Johnson to Tesla, spring 1926, Butler Library, Columbia University.
7. Letter, Tesla to Johnson, April 6, 1926, Butler Library, Columbia University.
8. Letter, Johnson to Tesla, 1926, Butler Library, Columbia University.

9. Colorado Springs *Gazette,* May 30, 1924, p. 1.
10. Microfilm letter, Tesla to Johnson, 1929, Library of Congress.

CHAPTER 25. *THE BIRTHDAY PARTIES*

1. Kaempffert, "Electrical Sorcerer."
2. "Tesla at 75," *Time,* July 20, 1931, pp. 27, 30; New York *Times,* July 5, 1931, II., p. 1; "Tesla, Electrical Wizard," Montreal *Herald,* July 10, 1931; "Father of Radio, 75," Detroit *News,* July 10, 1931; Kosta Kulišić, "Sedamdesetpetogodišnjica Nikole Tesle," *Politika,* Belgrade, July 11, 20, 21, 1931.
3. *Time,* July 20, 1931.
4. Ibid.
5. Ibid.
6. Ibid.
7. Ibid.
8. Nikola Tesla, "Our Future Motive Power," *Everyday Science and Mechanics,* December 1931, p. 26.
9. Letter, Prof. Warren Rice to author, September 5, 1980.
10. Dr. Gustave Kolischer, "Further Consideration of Diathermy and Malignancy," *Archives of Physical Therapy, X-Ray, Radium,* Vol. 13 (December 1932), pp. 780–81.

CHAPTER 26. *CORKS ON WATER*

1. Letter, Tesla to Viereck, April 7, 1934.
2. Letter, Tesla to Viereck, December 17, 1934.
3. Nikola Tesla, "A Machine to End War," *Liberty Magazine,* February 1935.
4. Ibid.
5. Ibid.
6. Ibid. See also, New York *Sun,* "Invents Peace Ray—Tesla Describes Beam of Destructive Energy," July 10, 1934; New York *Times,* "Tesla . . . Bares New 'Death Beam'" July 11, 1934; *Time,* "Tesla's Ray," July 23, 1934; New York *Herald Tribune,* July 11, 1934; New York *World Telegram,* July 10, 1937.
7. Letter, Tesla to New York *Times,* "Tribute to King Alexander," October 21, 1934, IV, p. 5.
8. Microfilm letter, Tesla to J. P. Morgan, November 29, 1934, Library of Congress.
9. Ibid.
10. Microfilm letter, Kintner to Tesla, April 5, 1934, Library of Congress.
11. Letter, Dr. Albert J. Phillips to author, February 10, 1979.
12. Ibid.

13. Microfilm letter, Johnson to Tesla, n.d., Library of Congress.
14. Microfilm letter, Johnson to Tesla, n.d., Library of Congress.
15. Microfilm letter, Johnson to Tesla, n.d. (mid-1930s), Library of Congress.
16. Microfilm letter, Johnson to Tesla, n.d., Library of Congress.
17. O'Neill, *Genius*, p. 313.
18. Microfilm letters, Westinghouse Company to Tesla, April 29, 1938, Library of Congress. See also: *Nikola Tesla. Spomenica povodom njegove 80 godisnjice. Livre commemoratif a l'occasion de son 80ème anniversaire. Gedenkbuch anlässlich seines 80sten Geburtstages.* Memorandum book on the occasion of his 80th birthday. Belgrade, *Priredilo i izdalo Društvo za podizanje Instituta Nikole Tesle;* Belgrade, *Edition de la Société pour la fondation de l'Institut Nikola Tesla,* 136, 519 pp. (tributes in original languages in which written).

CHAPTER 27. *COSMIC COMMUNION*

1. Nikola Tesla, unpublished paper, 1936, part appears in New York *Herald Tribune,* July 9, 1937, "Tesla Devises Vacuum Tube Atom-Smasher." Nikola Tesla, "German Cosmic Ray Theory Questioned," letter to New York *Herald Tribune,* March 3, 1935. See also, "Tesla, 79 . . . New Inventions," New York *Times,* July 7, 1935, II, p. 4.
2. "Tesla Has Plans to Signal Mars." New York *Sun,* July 12, 1937, p. 6; "Sending Messages to Planets," New York *Times,* July 11, 1937, II, p. 1; Detroit *News,* July 11, 1937.
3. Ibid, New York *Sun.*
4. *Science News,* April 30, 1977, Vol. III.
5. New York *Herald Tribune,* July 11, 1937; New York *Times,* July 11, 1937, II, p. 1.
6. William L. Laurence, New York *Times,* September 22, 1940, II, p. 7.

CHAPTER 28. *DEATH AND TRANSFIGURATION*

1. Peter II, King of Yugoslavia, *A King's Heritage* (New York, Putnam, 1954). See also, *The Balkans,* Life World Library (New York, Time, Inc., 1964). *Yugoslavia, Background Notes,* Department of State Publication 7773, Rev. February 1978, U.S. Government Printing Office. M. Djilas, *Memoir of a Revolutionary* (New York, Harcourt, Brace, Jovanovich, 1973).
2. New York *Times,* January 9, 1945, May 1, 1945, June 7, 1947.
3. Letter, Prof. Bogdan Raditsa to author, February 19, 1979.
4. Ibid.
5. Ibid.

6. Peter II, *A King's Heritage.*
7. Ibid.
8. Letter, Professor Raditsa to author.
9. Letters, Charles Hausler to Leland I. Anderson, April 12, July 16, 1979.
10. Report of Death, Office of Medical Examiner of the City of New York, January 8, 1943. See also, New York *Times,* January 8, 1943, p. 19; New York *Herald Tribune,* January 8, 1943, p. 18; New York *Telegram,* January 8, 1943, p. 36; New York *World,* January 8, 1943, p. 36; New York *Sun,* January 8, 1943; New York *Journal American,* January 8, 1943; New York *Times,* editorial, January 8, 1943, p. 12.
11. New York *Herald Tribune,* January 11, 1943.
12. Ibid.

CHAPTER 29. *THE MISSING PAPERS*

1. Formal agreement of International Electrotechnical Commission, Munich, June 29, July 7, 1956. See also Swezey, "Nikola Tesla."
2. FBI memorandum, New York Agent Foxworth to director, New York Bureau FBI, January 9, 1943.
3. FBI Memorandum, D. M. Ladd to E. A. Tamm, Washington, January 11, 1943.
4. Handwritten notation, Edward A. Tamm to D. M. Ladd, on memo of January 11, 1943.
5. Letter, Headquarters, Aeronautical Systems Division, Wright-Patterson Air Force Base, to author, July 30, 1980.

CHAPTER 30. *THE LEGACY*

1. Lambert von Ing. Binder, "*Portrat eines Technomagiers, Mensch und Schicksal* (Mankind and Destiny), January 15, 1952, Wien, Austria, pp. 3–5.
2. Robert Golka, "Project Tesla," *Radio-Electronics,* February 1981. See also, Charles Hillinger, "Lightning as New Energy Source, Los Angeles *Times,* April 29, 1979; San Francisco *Chronicle-Examiner,* May 20, 1979; R. K. Golka and R. W. Bass, "Tesla's Ball Lightning Theory," presented at Annual Controlled Fusion Theory Conference, May 4–6, 1977, San Diego, California; Reed Blake, "The Wizard of Wendover," *Mountain West,* November 1977, pp. 26–29.
3. P. L. Kapitza, "High Power Electronics," Russian periodical, *Uspekhi Fizicheskikh. Nauk,* Vol. 78 (November 2, 1962), pp. 181–265; *Life* magazine, June 16, 1961. See also, Jerzy R. Konieczny, "New Weapon 'X'—Ball Shaped Thunders (Globular Fireballs)," Polish periodical, *Wojskowy Przeglad Lotniczy,* November 2, 1963, pp. 72–75.

4. New York *Times,* October 10, 1975, p. 40 L. See also, Peter E. Glaser, "Solar Power from Satellites," *Physics Today,* February 1977, p. 30–37.
5. "Tesla Coil Almost Ready," St. Cloud (Minnesota) *Daily Times,* August 19, 1977, pp. 1, 19, 20. See also, R. J. Schadewald, "Power Could Be Dirt Cheap," Minneapolis *Star,* June 6, 1978; Yvonne Baskin, "Power from Earth," San Diego *Union,* July 23, 1980, pp. A-1, A-12.
6. Letter, James R. Wait to author, November 14, 1979.
7. James R. Wait, *IEEE Spectrum,* Vol. 16, No. 8 (August 1979), Book Reviews, p. 72.
8. Nikola Tesla, "Breaking Up Tomadoes," *Everyday Science and Mechanics,* December 1933, pp. 870–922. See also, Nikola Tesla, *Minutes of Edison Medal Meeting,* AIEE, May 26, 1917, Smithsonian Institution, p. 29.
9. Frederic Jueneman, "Innovative Notebook," *Industrial Research,* February 1974.
10. Letter, Lambert Dolphin to author, September 15, 1980.
11. Ibid.

POSTSCRIPT:
END OF A PAPER CHASE

Since the foregoing chapters were written and the proofs read and corrected, the disposition of Tesla's "missing" scientific papers, originally held by the Office of Alien Property, has become known to me.

I have learned that a substantial classified Tesla file is contained in the third of three libraries at a well-known defense research agency. One library is open to the public, the second is semi-restricted, and the third contains material seen only by members of the intelligence community. Tesla's ideas contained in the research papers so urgently requisitioned by military intelligence in 1947 have indeed continued to be of great interest.

When the Tesla Museum in Belgrade published *The Colorado Springs Notes, 1899–1900,* in 1978, intelligence officers at the said U.S. defense research agency at once made a careful comparison of both the Serbo-Croat and English editions against their classified files for that period in Tesla's life. What they found, I am reliably told, is that the Slavs had omitted mainly practical Teslian ideas that might prove patentable. His fundamental research in wave propagation, radio and power transmission, and ball lighting, however, was found to be substantially the same in the *Notes* as in the U.S. intelligence file.

But the file apparently contains much more than the *Notes.* It almost certainly contains the complete papers from which the Trump abstract, portions of which were quoted earlier, was derived. It undoubtedly contains the papers that the two young American engineers pored over night after night in a hotel room in the weeks just before Tesla's death. It probably contains the work papers which John J. O'Neill said were removed from his home by federal agents and which he was subsequently unable to trace.

What else may be in that intriguing file, I do not know. Nor do I withhold the name of the research agency possessing it merely to tantalize the reader; my only reason for doing so is that the U.S. government has deemed the material important to national security and has been at great pains to conceal its existence.

Today applications of scientific knowledge are being made at a dizzying rate. Shall we meet Nikola Tesla once again when we are farther down the road? I am sure of it.

INDEX

388

Index